Biological Neural Networks:Hierarchical Concept of Brain Function

Konstantin V. Baev

Birkhäuser
Boston • Basel • Berlin

Konstantin V. Baev
Department of Neurosurgery
Barrow Neurological Institute
St. Joseph's Hospital and Medical Center
Phoenix, AZ 85013-4496

Library of Congress Cataloging-in-Publication Data

Baev, K. V. (Konstantin Vasil'evich)
  Biological neural networks : hierarchical concept of brain
function / Konstantin V. Baev
    p.   cm.
    Includes bibliographical references and index.
    ISBN 0-8176-3859-8
    1. Neural networks (Neurobiology)   I. Title.
    [DNLM: 1. Brain--physiology.   2. Nerve Net--physiology.
  3. Automatism.   4. Learning--physiology.   5. Models, Neurological.
  WL 300 B1417b 1997]
  QP363.3.B34   1997
  573.8'6--dc21
  DNLM/DLC
    for Library of Congress   97-30734
                                                                CIP

QP
363
.3
.B34
1998

Printed on acid-free paper
© 1998 Birkhäuser Boston   *Birkhäuser*

ISBN 0-8176-3859-8
ISBN 3-7643-3859-8
Typeset by Northesatern Graphic Services,Hackensack, NJ.
Printed and bound by Edward Brothers, Ann Arbor, MI.
Printed in the U.S.A.

9 8 7 6 5 4 3 2 1

Contents

Preface

This book is devoted to a novel conceptual theoretical framework of neuroscience and is an attempt to show that we can postulate a very small number of assumptions and utilize their heuristics to explain a very large spectrum of brain phenomena. The major assumption made in this book is that inborn and acquired neural *automatisms* are generated according to the same functional principles. Accordingly, the principles that have been revealed experimentally to govern inborn motor automatisms, such as locomotion and scratching, are used to elucidate the nature of acquired or *learned* automatisms. This approach allowed me to apply the language of control theory to describe functions of biological neural networks. You, the reader, can judge the logic of the conclusions regarding brain phenomena that the book derives from these assumptions. If you find the argument flawless, one can call it common sense and consider that to be the best praise for a chain of logical conclusions.

For the sake of clarity, I have attempted to make this monograph as readable as possible. Special attention has been given to describing some of the concepts of optimal control theory in such a way that it will be understandable to a biologist or physician. I have also included plenty of illustrative examples and references designed to demonstrate the appropriateness and applicability of these conceptual theoretical notions for the neurosciences. However, this monograph is not entirely comprehensive for a few obvious reasons. First, a comprehensive text on such a broad topic would of necessity be voluminous, clumsy, and very unreadable. Consequently, those scientists who do not find references to their publications should not think that they have been deliberately neglected. Second, a conceptual description such as this one does not require a comprehensive citation of the scientific literature.

It is necessary to mention that several scientific trends of the last two decades have seriously influenced the development of concepts proposed in this book. Progress in the field of neurocomputing in the 1980s has helped me to refine the notion of computation and its implications for understanding the nervous system. In the 1980s, neurocomputing became very popular as a result of the work of Hopfield (Hopfield 1984, 1985; Hopfield and Tank 1985), and it was during this time that I first encountered a reference to Kolmogorov's theorem in the neurocomputing literature. This astounding theorem describes the very nature of network computational principles. I believe that neurocomputing was the catalyst for the creation of the discipline of computational neuroscience, and in 1994 a journal with the corresponding title was founded. The fundamental notion in neurocomputing and computational neuroscience is that, from a mathematical perspective, an artificial or a biological neural network should be capable of performing approximations of various mathematical functions. During the last two decades, the functional approach based on control theory has found serious application in fields relating to motor control, and in explaining the highest brain functions (see, for example, Arbib 1987, 1995; Grossberg and Kuperstein 1989; Grossberg and Merrill 1996). In this connection, it is also necessary to mention Anokhin (1974), Bernstein (1966, 1967), and von Holst (1954), who wrote about the importance of the functional approach to neurobiological problems many years ago.

This book is written for neurobiologists, neurosurgeons, and neurologists, and also for physicists and specialists in technical fields pursuing the design of artificial network computers based on the principles of brain function. But those who will benefit most from this monograph are undergraduate and graduate students interested in pursuing careers in the neurosciences and related disciplines. It is my hope and desire that this text will enable them to understand that the reflex theory stage in neuroscience is over, and that future neuroscience will evolve into a highly technical discipline. With this in mind, there is still enough time for them to modify their curriculum by completing the necessary courses in mathematics, physics and control theory. This will make them more competitive in their future scientific pursuits and help them appreciate the real value of the power of reasoning, which will save them a lot of time and energy in their future research careers.

Konstantin V. Baev

Acknowledgments

This is a perfect opportunity to thank all of those individuals and organizations that made this monograph possible. My primary thanks go to my former colleagues V. Berezovskii, N. Chub, A. Degtyarenko, V. Esipenko, T. Kebkalo, S. Kertzer, I. Melnik, K. Rusin, B. Safronov, L. Savos'kina, Y. Shimansky, and T. Zavadskaya. They contributed substantially to the experimental and theoretical results that became the basis for this book. I must also acknowledge the invaluable contributions of the technical personnel of my former department, G. Duchenko, I. Gerasimenko, A. Matzeluch, A.

Petrashenko, Y. Strelkov, and O. Starova, whose support of my experimental research was crucial. I hope that this book will bring back good memories for my former colleagues and remind them of one of the most scientifically fruitful times of our lives.

Writing a book involves a major commitment of time and energy, and I am very grateful to Dr. Robert F. Spetzler, Chief of the Division of Neurological Surgery and Director of the Barrow Neurological Institute; Dr. Abraham N. Lieberman, Chief of the Movement Disorders Section in the Division of Neurology at Barrow Neurological Institute and Director of the Arizona Branch of the National Parkinson's Foundation; and the National Parkinson's Foundation for the financial support necessary for writing this monograph.

The love, understanding, and patience of my wife, Tatiana, and my son Denis are gratefully appreciated. The early influences of my mother and father are also reflected in this book and are acknowledged here.

Finally, I offer my sincerest gratitude and appreciation to the editor of this monograph, Alla Margolina-Litvin, and to the reviewers of this book, Dr. Karl A. Greene, Chief of the Division of Neurosurgery and Director of the Institute for Neuroscience at Conemaugh Memorial Medical Center in Johnstown, Pennsylvania; and Dr. Alex Meystel, Professor of Electrical and Computer Engineering, Drexel University, for fruitful discussions and comments that were extremely valuable and supportive. Their help allowed me to improve the composition and style of this monograph. If the text of this monograph is read with ease, it is to a significant degree a reflection of their tireless efforts.

Foreword: Hierarchies of Nervous System

Cybernetics was embraced by the biological sciences and the neuroscience long time ago. The real advantage of this can be achieved if they made a step further: toward the discipline and structured analysis characteristic of control science. In his new book, K. Baev made this step.

As a specialist in control science and as a researcher in the area of intelligent systems, I was fascinated with the way of using control theory for interpreting the most difficult problems of brain functioning. Definitely, the book reduces the gap between neuroscience and control science to the degree that both will be able to use each other as a research tool, not just a source of metaphors.

1. What is this book about?

The book presents a novel view upon the nervous system and the brain as a part of it. Most of the previous literature considered control diagrams and cybernetic terminology to be a kind of knowledge organizer. They allow for an efficient presentation of knowledge in the area of formidable complexity. K. Baev went further: for him control theory is a source of insight and a tool for explaining away the physiological phenomena of the brain.

This book contains the first attempt to interpret processes of motion control in a biological organism driven by its nervous system as an optimal controller which presents the motor reflexes as the deliberately planned motion.

How can it be done in a system which contains many thousands of distributed "control loops"? This can be done only by using a powerful tool: constructing nested hierarchies of functional loops. K. Baev is doing this elegantly and persuasively.

This book is about the nervous system, a system which transforms knowledge. The shortest definition of "nervous system" is the semiotic one: it is a system which receives and transforms knowledge of the world so as to figure out a way to change this knowledge to benefit the carrier of this nervous system. To help understand the nervous system, it would be beneficial to understand what *knowledge* is. Knowledge is a relational network of symbols in which relations are also symbols.

What makes this peculiar network *knowledge*? A few things should be known about it. Each object of reality, or an event of the external world, or an actually existing system, can be put in correspondence with a symbol of the network which represent one of these entities. A set of such symbols can be generalized into a new entity. This entity will belong to another relational network, in which all entities are obtained as a result of this generalization.

The art of generalization can be applied to the entities of this second network, and a third network can be built. These new networks contain units of knowledge which are more and more generalized and have fewer details about the larger number of objects of reality. Each consecutive network will be called a "level of granularity", or a "level of resolution."

A nervous system cannot deal with the whole network at a particular level of resolution; it selects a "scope of attention." At each moment of time, the nervous system processes knowledge arriving externally and internally. In the meantime, the amount of knowledge the nervous system processes is limited. No more knowledge is processed than that which goes into our scope of attention. Within our scope of attention, we cannot distinguish knowledge "finer" than the input resolution (smallest distinguishable unit) our nervous system is capable of handling.

We are not able to deal with knowledge in a different manner: too much work must be done if one wants to avoid generalization. It turns out that our nervous system is built specifically to enable us to deal with the processes of generalization, and with the opposite processes which are called instantiations. How does the nervous system do this job? By the means and tools of neurophysiology. The latter contains all clues about dealing with knowledge for survival.

K. Baev's book does for neurophysiology what S. Grossberg's papers on adaptive resonance theory have done for neurocognition. It opens a new venue for interpreting the nervous system and brain.

When a critical amount of information is gathered in any branch of science, this usually means that time has come for generalization, and K. Baev undertook this effort. In this book, the author integrates an enormous amount of sources, and encapsulates views from these sources into a structure which is surprisingly clear and object-oriented[1]. Like other object-oriented systems, it has the advantages that come with *multiple granularities* and *nestedness*.

---

[1]For explanation of terms here and throughout the text, see **Glossary** in the end of this Foreword.

Both multiple granularities and nestedness are new phenomena for a scientific discourse on neurophysiology, as well as many of the disciplines, although the advantages of these phenomena were understood a long time ago. K. Baev approaches and applies them boldly and resourcefully. Using structural hierarchies and organizing neurobiological knowledge in an object-oriented way gives K. Baev an immense power of interpretation. However, it also brings in a set of problems in tuning up the reader for this uncommon, even singular, mindset.

Many researchers are used to criticizing any hierarchical knowledge organization because of the real and imaginary losses it might inflict. Especially critical are the bearers and carriers of high resolution knowledge: the more generalized the knowledge is, the more details seem to disappear within generalized entities. The situation is worse when the knowledge is multidisciplinary, and the recipients of knowledge restore details depending on their background.

The innovative conceptual paradigm in which K. Baev builds his *theory of multigranular automatisms* grows within a multi-disciplinary and thus, potentially contentious atmosphere. Before the reader gets too involved, let us clarify several important issues which can reconcile differences in the background and illuminate obscurities within this new and unexplored paradigm.

I will discuss concepts from various disciplines which will help to demonstrate how Baev's hierarchy of automatisms emerged. The following issues will be addressed.

- The evolution of our skills in symbols organization leads us to the concept of automatism.

- Analysis of learning processes explains the different granularities of automatisms.

- Finally, all of them are put together into a hierarchy of nervous system by using results of automated control theory.

2. Sign, Schema, Semiotics

From signs to architectures. K. Baev faces a formidable challenge: to integrate a tremendous amount of essentially multidisciplinary knowledge, which has different degrees of generalization. To propose his theory of multigranular automatisms, Baev applies an innovative conceptual approach, and the tools of building models for incompletely known systems of high complexity. These tools are used in control theory for building models of control.

The following components are used for building models: signs, schemata, symbols, concepts or categories, and architectures. A sign is associated both with an interpretation of concepts behind the sign, and with a real object and/or event. Symbols represent the next level of abstraction; they are labels associated with grouping things, qualities, events, actions, etc. A schema is a sign for rules (or implications) linking cause and effects it implies. All these components can and should be verified by a procedure of symbol grounding.

The need for signs, symbols and symbol grounding invokes the need for semiotics. The essence of the semiotic approach is in introducing a system of signs which is consistent at a particular scale, or level of resolution (granularity, scale). It allows for construction of an elementary loop of functioning *(or control)*. We will consider an elementary loop of functioning as shown in Figure I. (Figure 22 from Baev's book can be easily identified with Figure I).

This loop works as follows. The World within the scope of attention is registered by Sensors and enters Perception where the primary organization of the input information is performed. After that, the knowledge base is accumulated, where the World Model is formed. Based upon the World Model a set of commands is generated and the process of Actuation starts. The results of Actuation change the World, and Sensors perceive the changes. Functioning of Perception, World Model and Command Generation is presented in more detail in Figure II, in a discussion of the learning process associated with each elementary loop of functioning (called semiosis). Learning in a loop of semiosis produces such signs as S-R (stimulus-reflex) rules, or as they are sometimes called, *schemata*. They are, in fact, the signs created as a result of learning.

Automatism of schema is a generalization for a *percept–action* couple. In Baev's theory, the concept of automatism is a further development of the idea of schema at a particular level of resolution (granulation). This is a schema supplemented by the description of an elementary functioning loop in which the phenomenon of automatism is produced. Clusters of higher resolution units are unified by virtue of generalization (or *gestalt*). The idea

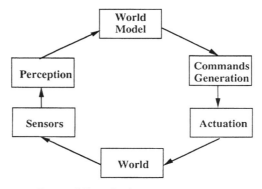

Figure I.—Elementary Loop of Functioning

of unity was extremely productive in the area of perception. It becomes as useful in the area of motor control by forming the couple of "percept-action" (p–a). M. Arbib proposed to use this couple as a primitive unit for solving problems of motion control: "We owe to the Russian school founded by Bernstein the general strategy which views the control of movement in terms of selecting one of a relatively short list of modes of activity, and then within each mode specifying the few parameters required to tune the movement. . . . we will use the term *motor schema*" [1].

A precursor to Baev's aggregation of automatisms of higher resolution into more general automatisms of lower resolution can be found in the aggregation of schemata. Motion generating schemata can be considered a multiresolutional language for synthesizing the behavior of a system, i.e., for "sequencing and coordinating such motor schemata" [1]. This language is used to describe automatisms by using its vocabulary, grammar, axioms, tools of generalization, focusing attention, and combinatorial search. It also has a context in which the generated statements can be interpreted. The multiresolutional structure of this language minimizes the complexity of behavior-generating processes. The words of the vocabulary emerge as a result of learning, which is performed through an intentional synthesis of alternatives and search among them, or through involuntary generation and testing of alternatives as a part of the experience of functioning. Groups of kindred schemata are generalized into a single schema of lower resolution. This substantially simplifies the related subsequent processes of storage and retrieval and reduces complexity.

Most of the high resolution schemata in living creatures are their reflexes. Taxis create another group of learned patterns of behavior which can be considered schemata of a higher level of generalization (lower level of resolution) [2, 3]. Instincts are the most general schemata [4–6]. Some researchers arrive at the concept of hierarchical organization of behavior [7, 8]. Analysis of examples, e.g., for the herring gull in [7], demonstrates a multiresolutional organization of schemata. This correlates with the hierarchical organization of sensory categories (see Chapter 6 of "The Geometric Module in the Rat" in [9]). Decomposition of schemata into subschemata is emphasized in [10]. An illustrative process of "categorization of movements" is presented in [11].

The concept of automatism is a further development of the idea of schema. This is a schema supplemented by the description of an elementary loop of functioning in which the phenomenon of automatism is produced.

3. Evolution of the Notion of Automatism

Learning Automata. Sets of rules, or schemata, as a tool of representing systems can be put in the framework of automata theory. This theory requires that input and output languages be defined. Then, the concept of "state" is introduced. Note that the automata theory does not contemplate such realistic things as the "outside world". The latter arrives to an automaton as a set of messages encoded in the input language. However, if one tries to apply this theory to the central nervous system (CNS), the symbols and/or codes should be equipped with interpretations. Input messages are delivered through sensors. State is a representation of the outside world within the automaton. After this, the transition and output functions can be determined as a product of the functioning of the CNS. Certainly, our actions are the output statements.

K. Baev's book is a pioneering work because it puts together the areas of neurobiology or neuropsychology with the areas of automatic learning control and learning control theory. Evolutionary development of the higher nervous system is equivalent to the development of learning how to act—this is what automatisms are all about. The author comments on reflex theory that it was easily accepted because of its conceptual simplicity. Yet, when reflexes are clustered and these clusters become entities, in which learning works at lower resolution, the conceptual structure requires from the reader a little bit of insight and imagination. This is why the concept of *automatism* is introduced as fundamental component of the hierarchy of nested loops in the CNS.

Automatism emerges as a result of learning. The term "automatism" is rare in books on neurophysiology or neuropsychology. In dictionaries, it is interpreted as follows: the state or quality of being automatic; automatic mechanical action; the theory that the body is a machine whose functions are accompanied but not controlled by consciousness; the involuntary functioning of an organ or other body structure that is not under conscious control, such as the beating of the heart or the dilation of the pupil of the eye; the reflex action of a body part; mechanical, seemingly aimless behavior characteristic of various mental disorders.

Practical applications of this term are frequent in engineering. It is not unusual in medical science. For example, *automatism* is used to characterize aimless behavior which seems to be not directed and unconscious, with no conscious knowledge processing: it is seen in psychomotor epilepsy, catatonic schizophrenia, psychogenic fugue, and other conditions. A phenomenon is known called *traumatic automatism* which is characterized by performing complicated motions (e.g. playing football) totally automatically. Temporal lobe epilepsies are often characterized by automatisms such as the repetitive opening and closing of a door.

K. Baev has introduced this term to demonstrate the commonality of automatic control processes at different levels of motion control in the CNS. His goal is to demonstrate that all hierarchical levels of the nervous system are built according to the same functional principles, and each level is a learning system which generates its own automatism [12].

The tradition of behavioral science in the US is to talk about stereotyped responses which are usually separated into four categories: a) unorganized or poorly organized responses, which describe processes in organisms lacking a nervous system, b) reflex movements of a particular part of an organism as a result of the existence of a reflex arc (an open loop pair of receptor and affector is usually mentioned), c) reflex-like activity of an entire organism, d) instinct.

"Reflex-like activities" is a label for complicated phenomena. Reflex-like activities of entire organisms may be unoriented or oriented. Unoriented responses include kineses— undirected speeding or slowing of the rate of locomotion or frequency of change from rest to movement (orthokinesis) or of frequency or degree of turning of the whole animal (klinokinesis). Oriented reflex activities include tropisms, taxis, and orientations at an angle. Their mechanisms are rarely discussed. The tradition of behaviorism is to talk about them as "reflex actions of entire organisms." Their classification is done based upon external features, e.g., tropotaxis serves for orientation based upon some need, while telotaxis is an orientation toward light.

Instinctive behavior is considered to be an unlearned, rather hereditary property of an organism (like cleaning, grooming, acting dead, taking flight and so on.) Behaviorists analyze the complexity of patterns of instinctive behavior, their adaptivity, stability, etc. K. Baev is interested in discovering their inner mechanisms. He unifies all of them under the title of "automatisms" and demonstrates that they have similar control architectures. However, they belong to different levels of granularity (resolution) in the CNS. Stereotyped responses of all types can only emerge if a particular mechanism of learning is assumed. Then, the gigantic body of information consisting of seemingly unrelated units becomes well organized and explainable.

From reflexes and rules to programs. Learning is a development of automatisms, (in Baev's sense). As new environments persist and new experiences perpetuate, new rules emerge as a result of generalization upon repetitive hypotheses generation (see Figure II). Learning on a large scale in time, such as an evolution of the nervous system of the particular species, can be also considered an evolution of the automatisms.

World Model from Figure I is acquired via learning process. Understanding of this process is important for us because World Model can be considered a collection (a hierarchy) of automatisms learned via a process of collecting and generalizing experiences. The process of experiences acquisition and their transformation into rules is demonstrated in Figure II.

The system functions as follows:

1. Experiences are recorded in the form of associations which incorporate four components: state, action, change in the state, value. The system of

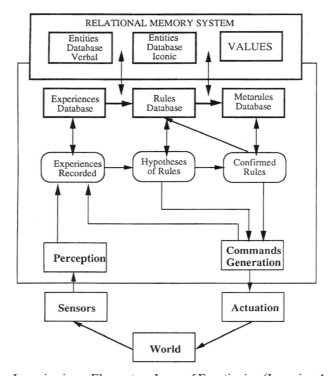

Figure II.—Learning in an Elementary Loop of Functioning (Learning Automaton)

storage allows for clustering these associations by similarity in each of the elements of the association.

2. The clusters are generalized in the form of hypotheses of future rules.
3. Elements of the rules are organized in the form of concepts.
4. After confirming the hypothesis, the latter are assigned a higher value of preference.
5. As the number of rules increases, the same procedure of clustering is applied to them. Metarules are obtained; metaconcepts are extracted from them.

   Once the process of learning starts, it cannot stop since new rules generate new experiences. The new experiences give basis for new generalizations, and then new levels of generality emerge.
6. Multiple repetition of steps 1 through 4 leads to the formation of hierarchies of rules, metarules, concepts.

Algorithms of learning equivalent to the one shown in Figure II have been tested in a variety of systems of machine learning [13, 14]. Various types of hardware can be used for implementation of this algorithm of learning. If the hardware of the system is our nervous system, then instead of the hierarchy of rules, we obtain the hierarchy of automatisms which allow for control of the system. As a result of this development, the loop shown in Figure I is transformed into the nested multiresolutional loop shown in Figure III.

4. Controllers Within Nervous Systems

Consistency and Intelligence of Automatic Control. K. Baev persistently introduces control theory and corresponding terminology in his discourse. His striving for consistency is fascinating. It results in a theory satisfying the most demanding scientific standards.

In the area of neurobiological architectures of control, we will always deal with a contradiction between the following factors:

a) our need to formulate specifications of the system, conditions of the test, and premises of the derivation with a level of scientific rigor typical for the control community,

b) the impossibility to provide sufficient statistics for a proof, interference of personal views and interpretations, etc.

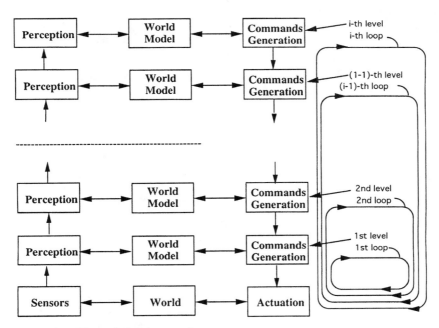

Figure III. —Hierarchy of Automatisms

It is not easy to discuss the mechanisms of how the nervous system functions as a whole: too much common-sense reasoning would be involved, too much of a multidisciplinary blend is required in which no authority can approve the line of reasoning. The most intimate subtleties of the CNS functioning often seem to be illogical: how can they be explained within the framework of conventional logic? One can see this in any discussion concerning the definition of *intelligence*. It may happen that no scientific unity of results will ever be achieved in this area. Nevertheless, Baev persistently generalizes the nervous system and its parts into a set of automatisms. When the boxes of generalized subsystems emerge, it is control theory's turn to speak. Somewhere at the top of an hierarchy, intelligence's turn will come too: it so happened in humans' systems.

Control structures. In Figure IV, a single level architecture of control is demonstrated in the most general form. Here, PLANT is the system to be controlled (e.g. muscles and the external objects they act upon), COMMAND GENERATOR is the feedforward controller (e.g., a subsystem of the nervous system) which generates temporal strings of commands by using its knowledge of the model of PLANT (by inverting the desired motion, or using a look-up table).

It takes a computational effort to find the inverse of the desired trajectory (or to find a set of commands, or strings of commands, which will generate the desired trajectory at the output). In order to simplify the feedforward controller, instead of inverting, we can use a library of stored command strings which can be considered elementary automatisms. This is equivalent to the storage of rules shown in Figure II. If our knowledge of PLANT is perfect (which never happens), no other control is required, the feedforward controller is sufficient, and our control system is the Open Loop Controller (OLC) which is shown in Figure V in bold lines.

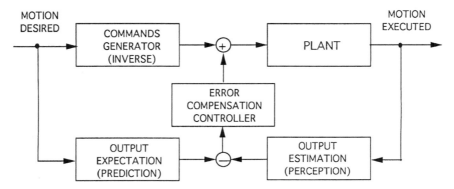

Figure IV.—Control System at a Level

Since the real PLANT is different from our knowledge of it, the expected and true output movements will differ. The difference must be computed and compensated by the Feedback (Closed Loop) controller which is shown in Figure V in dashed lines. It is imperative to properly estimate the output motion using sensor information. Also, we should not forget about the delay Δt between the cause and effect: the output appears a little later than the control command is issued. Thus, it would be useful to properly compute the *predicted* output. As the command is issued at time t, a prediction should be made about the output value at time (t + Δt.) As S. Grossberg reminds "Perceptions are matched against expectations" [15]. The feedforward-feedback (OLC-CLC) architecture of single level control is widespread in the literature. M. Arbib uses it for analysis of the CNS (it is given in [16] with a reference to his book of 1981).

Levels of granularity differ in their frequency. The error (the difference between the expected and real signals) has interesting characteristics. It is usually much smaller in magnitude than the desired signal. Its spectral density (the package of signal frequencies it contains) is shifted toward higher values on a frequency scale. Clearly, instead of using the model of PLANT which contains full information required for all frequencies, one can use two models together: one simplified model for the lower frequencies of the desired signal and another simplified model for the frequencies of higher bandwidth.

If this is true, the sampling frequency of the compensation loop should be higher than the sampling frequency of the feedforward channel. On the other hand, the spatial resolution of the compensation loop should be higher than the spatial resolution of the feedforward channel. Thus, a conjecture can be proposed that the compensation process (for the feedforward channel) belongs to the level of resolution higher than the resolution of the feedforward channel. At this level, the compensation commands for level 1 can be consid-

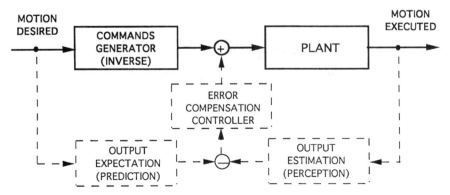

Figure V.—Open Loop Controller, OLC (Feedforward) shown in bold lines versus Closed Loop Controller, CLC (Feedback) shown in dashed lines

ered a feedforward control command for level 2. Our analysis for levels 1 and 2 can be recursively repeated for levels 2 and 3, and a multiresolutional (multigranular, multiscale) hierarchy of control is obtained (Figure VI).

Control Hierarchies. There are other ways of introducing control hierarchies. One of them is associated with the top-down task decomposition. For example, each task can be broken down into its components as described in [17]. These components are always spatially smaller and temporally shorter.

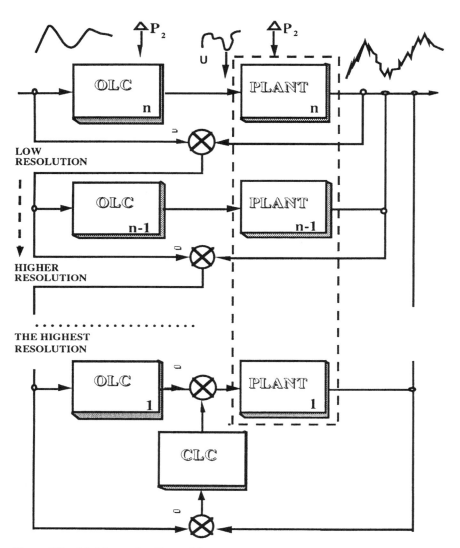

Figure VI.—Multigranular Control System

Thus, the temporal and spatial resolutions of the elementary actions (automatisms) will grow top down from level to level of the control hierarchy.

Control hierarchies of nervous system were anticipated a long time ago [18]. The rationale for them was clear: task decomposition, gradual focusing of attention, and increase of resolution were a powerful source of reducing the complexity of processing [17]. This hypothesis was later confirmed quantitatively [19]. One Section in [20] is called "There Are Three Levels in the Hierarchy of Motor Control." A hierarchy of motor control system is described in [21]. One can see that α- and γ- motoneurons are above muscles, spinal interneurons are above motoneurons, brainstem is above spinal interneurons, motor cortex with premotor cortex are above brainstem. At the top, we arrive at the subcortical areas and association cortex. The latter are connected to thalamus, basal ganglia, and cerebellum which participate in creation of automatisms together with the upper levels of the hierarchy. Although, the top-down hierarchy in control distribution is there, no hierarchy in sensory processing is admitted. Only feedforward control connections are shown as if no feedback exist from each lower level to an adjacent level above (see [21], p. 413).

K. Baev is the first who consistently demonstrates that control systems within the nervous system and brain can be represented at all levels *in a uniform way*.

Deliberative Planning: A Challenge for the Future. K. Baev recognizes a formidable fact which creates a difference between the standard (Wiener's) cybernetics and the cybernetics of the mth order: *systems are driven by control loops of various granularity and systems of coarse granularity drive and incorporate system of finer granularity*. Moreover, K. Baev represents and explains how these multiple nested control loops emerge in the nervous system. However, so far we have only discussed control loops capable of generating *automatisms*, not *deliberative motion*.

There exists a tradition of explaining control of deliberative motion as a part of the hierarchical functioning of the nervous system. J. M. Fuster describes the hierarchical theory of nervous system as follows : "In general, elaborate and deliberate actions are represented in the cerebral cortex, simple and automatic actions in subcortical structures, the cerebellum, the brain stem, and the spinal cord. Less widely accepted is the implication that movement is hierarchically controlled from the top down"[22].

Multiple intuitive statements from [23–26] support this statement. However, recent publications [27, 28] differ on this issue. In [27], the concept of hierarchical control is muted. An architecture is proposed which is vaguely hierarchical. It is far from the explicit four levels proclaimed in [28] where the levels of motor neurons, α-neurons, brain stem, cerebellum, precortical and cortical areas are drawn with the clarity of a blueprint for the structure of a manufacturing company.

Obviously, the control assignments which come from the cortex as well as from all levels top-down are feedforward control commands in the sense of Figures IV and V. It is also clear that the bottom-up signals are feedback commands. We do not know with full certainty whether they carry information about the output, or about the discrepancy between the expected motion and the actual motion. We can only conjecture about the process of decomposing more general motion assignments into the hierarchy of automatisms.

In his Sherrington lecture of 1982, C. G. Phillips attempts to move from analysis of the "most automatic" to the "least automatic" movements [29]. This tendency is now pervasive: this is where automatisms merge with intelligence! A review of the present state of the art in the area of deliberative motion control can be found in [27]. S. Grossberg's recent models are oriented toward a connectionist hypothesis of voluntary motion [30].

K. Baev's model of the multigranular control system allows us to explore the possibility of integrating multiple existing theories into a unified model. This model allows for synthesizing voluntary motion out of "standardized" components. The role of these components is played by Baev's automatisms.

Baev's model raises one more conceptual and even semiotic issue: can this model incorporate both reactive and active behaviors, and how? It is true that the stored decision table of automatisms contains reactive rules, and reflexive actions generated by them always emerge as a reaction to some external stimulus. Apparently it was sufficient for survival because all possible capabilities of learning were involved in the development of this table of automatisms. However, the difference between *active* and *reactive* becomes blurred since what we call *active* can simultaneously be considered *reactive to the anticipation*.

Pure reactive behavior with no prediction is known to be insufficient for survival. This is why prediction and other forms of deliberation are parts of the functioning of the nervous system in its motion control activities.

Optimization. One of the key concepts in Baev's book is optimization. It is clear for him that if an automatism has developed then some "cost" was minimized, and this "cost" is the one which is required to be watched for survival. This changes a control problem: instead of the goal being to minimize the deviation from the prescribed trajectory, we have to minimize a cost-functional while keeping the deviation within some bounds.

Then, the functioning of ANS can be treated as the functioning of a multiresolutional nested controller in which the same search procedure of finding feedforward control (planning) is executed consecutively top down at several resolution levels within a subset of attention which becomes more narrow from one level to another. This concept does not neglect dynamics: the search is performed in a state space. Thus, an opportunity has emerged to solve the problem of optimum planning, e.g., minimizing the time of operation. The theory of multiresolutional hierarchical nested control as presented in [31, 32] matches the expectations of Baev's theory [12].

4. The Architecture of the Nervous System (ANS)

Hierarchical ANS. Baev's theory boils down to outlining the multiresolutional hierarchy of control as an ANS. This view should be obvious to everyone who was ever involved in the analysis of large complex systems. A multigranular hierarchy provides for computational efficiency, and allows for drastic complexity reduction [19].

This concept is not unanimously accepted by the scientific community. The opposing voices stem from either prejudices and political preferences ("hierarchies cannot be efficient: look at the bureaucratic hierarchy of administration, or the army hierarchy") or from a habit of the scientists to enjoy the analysis of systems at one level of resolution: the bottom level with the highest resolution available (see for example [33]). A more balanced approach is reflected in [34]. Churchland and Sejnowski do not reject hierarchical architecture completely but they do not embrace it as a major principle of organization of the CNS either.

Baev's theory of control hierarchy as ANS is conclusive. Evolution, with its struggle for existence, appears to construct creatures and systems whose controllers are hierarchical. ANS must be hierarchical because to survive one must optimize. To optimize one must organize perceptions, knowledge, and control in a multigranular hierarchical manner. However, it does not mean that ANS hierarchies look nice and crisp (like Figure V of this Foreword.) They often look clumsy, the boundaries between levels are fuzzy, and one should consider many research fields simultaneously before the hierarchies can be visualized and analyzed. Such view of ANS can be found in [31, 32, 35]. All of these architectures become hierarchies as soon as large amounts of data are to be organized, or large amount of computations are to be executed. The hierarchical structure of ANS is a tool of reducing complexity. Indeed, by using a hierarchical ANS we can make the complexity of an NP-complete problem manageable by using the consecutive functioning of hierarchical levels.

Control Tools of Hierarchical ANS: Multiresolutional Refinement and Search. Search in a state space [32] is done by synthesizing alternatives of motion and scanning the set of available alternatives. It is done in the following manner. A number of points is highlighted in the state space. They are contemplated as possible intermediate states of the future motion trajectory. Then, combinations of these points (strings of them) are constructed and compared. One of the strings is to be selected which has the highest degree of the desirable property of motion (e.g., minimum time). This string is considered to be the solution (the trajectory to be followed, the plan for feedforward control.)

The vicinity of the solution is considered at the adjacent higher resolution level where the search is repeated again, however, it is executed only within the vicinity. Thus, we receive a refined trajectory and send it again to the adjacent higher level of resolution. It is not a method of centralized control; the consecutive refinement is a general technique which is also applicable for decentralized solutions. A triplet of operations of *focusing attention*, *combinatorial search*, and *grouping* is performed consecutively with increasing resolution at each repetition of the triplet until the resolution of the level is equal to the accuracy of the decision required. This triplet is characteristic for all algorithms implementable in the controllers for intelligent systems.

Indeed, the process of deliberative decision making (planning) starts with *focusing attention,* which is a selection of the initial map with its boundaries. *Combinatorial search* is performed as a procedure of choosing minimum cost string from the multiplicity of all possible strings formed out of elementary units of space of resolution. *Grouping* is the construction of an envelope around the vicinity of the minimum cost string. These three procedures together amount to generalization: they produce a new entity. This entity is submitted to the next level of resolution where the next cycle of computation starts.

5. The Roadmap to the Future of Neurobiology

Multidisciplinarity. Baev's theory crosses the borders of many disciplines. This is where it gets its power from, and this is where the major difficulty arises for using this theory in practice. It attracts our attention to the fact that there is an invariance in the learning processes at all levels of resolution of the nervous system. This invariance is reflected in the signs created at each level of the semiotic system. In the case of the nervous system, this invariance is called *automatism.*

This book requires from biologists to be ready to think and model in terms of control theory. It requires the control community to get involved with the weakly organized bulk of knowledge on the Central Nervous System. This book is actually defying the pigeon-hole principle as the major principle of doing science.

Significance of the theory of automatisms. We should understand automatisms as a basis for making decisions about the future, when reusable solutions exist. Automatisms are good if they are assumed to be just a vocabulary for synthesis of more general automatisms. Here we arrive at the grey area: generalizing automatisms until they become a deliberative movement (maybe, even until they become an original solution, an insight, a discovery) . . . Can this happen? From the educational point of view, this is the scenario which mimics the work of our brain. Baev undertakes a simplifcation of the model by jointly applying the elementary functioning loops (see Figure I) and the multiresolutional view (Figure III), which is the right thing to do in teaching the "art of creative simplification."

Unsolved Problems. A book may be judged by the number of unsolved problem it helps to discover. The greatness of a book can be in its ability to uncover paths leading to the wide open spaces from which new and surprisingly unfamiliar vistas can open to us.

I have mentioned a few of the new problems that I saw. They include the following questions:

- Are voluntary motions synthesized from involuntary ones?

- Does general planning use the vocabulary of elementary automatisms?

- Is there an element of search at the level of control, or are the process of synthesis and its results predetermined?

- Do we process current control information in terms of discrepancies between the real and expected motion, or as a complete information about the motion of interest?

This list can go on, but I would like to delegate the pleasure of doing this to the reader.

References

1. M. A. Arbib,"Modeling Neural Mechanisms of Visuomotor Coordination in Frog and Toad", Chapter 21 in Eds. S. Amari, M. A. Arbib, *Competition and Cooperation in Neural Nets*, Proc. Kyoto, 1982, Springer-Verlag, Berlin, 1982
2. J. P. Scott, *Animal Behavior*, The University of Chicago Press, 1958
3. P. Marler, W. Hamilton, *Mechanisms of Animal Behavior*, Wiley, NY 1966
4. N. Tinbergen, *The Study of Instinct*, The Clarendon Press, Oxford 1951
5. W. Thorpe, *Learning and Instinct in Animals*, 1963
6. K. Lorenz, *Evolution and Modification of Behavior*, London, 1965
7. N. Tinbergen, "The Hierarchical Organization of Nervous Mechanisms Underlying Instinctive Behavior, Symp. Soc. Exp. Biol., v. 4, 1950
8. G. P. Baerends, "On Drive, Conflict and Instinct, and the Functional Organization of Behavior, Eds. M. Corner, D. Swaab, Progress in Brain Re. Vol. 45, *Perspectives in Brain Research*, Elsevier, Amsterdam, 1976
9. C. R. Gallistel, *The Organization of Learning*, The MIT Press, Cambridge, MA 1990
10. D. E. Rummelhart, "Schemata: the building blocks of cognition", in Eds. R. Spiro, B. Bruce, and W. Brewer, *Theoretical Issues in Reading Comprehension*, Erlbaum, Hillsdale, NJ 1980
11. C. G. Phillips, *Movements of the Hand*, Liverpool University Press, 1986

12. K. V. Baev, "Highest Level Automatisms in the Nervous System: A Theory of Functional Principles Underlying the Highest Forms of Brain Function", Progress in Neurobiology, Vol. 51, 1977

13. A. Meystel, "Learning Algorithms Generating Multigranular Hierarchies", Proc. of the Workshop on Mathematical Hierarchies and Biology, DIMACS Center, American Mathematical Society, 1997

14. J. Albus, A. Lacaze, A. Meystel, "Autonomous Learning via Nested Clustering", Proc. of the 34th IEEE Conference on Decision and Control, Vol.3, New Orleans LA, 1995

15. S. Grossberg

16. M. A. Arbib, The Metaphorical Brain 2, Wiley, NY 1989

17. J. Albus, Brains, Behavior, and Robotics, BYTE/McGraw-Hill, 1981

18. J. Szentagotai, M. A. Arbib, Conceptual Models of Neural Organization, A report based on an NRP Session, October 1-3, 1972, Yvonne M. Homsy, NPR Writer-Editor, Boston, 1974

19. Y. Maximov, A. Meystel, "Optimum Design of Multiresolutional Hierarchical Control Systems", Proc. of the 7-th IEEE Int'l Symposium on Intelligent Control, Glasgow, GB 1992

20. C. Chez, "The Control of Movement", in Eds. E. R. Kandel, J. H. Schwartz, T. M. Jessel, Principles of Neural Science, Appleton & Lange, Norwalk, CT, 1997

21. B. Pansky, D. J. Allen, G. Colin Budd, Review of Neuroscience, Macmillan, New York, 1988

22. J. M. Fuster, Memory in the cerebral cortex: An Empirical Approach to Neural Networks in the Human and Nonhuman Primate, The MIT Press, Cambridge MA, 1995

23. J. H. Jackson, "On affections of speech from desease of the brain," Brain 38: 107-174, 1915

24. N. Bernstein, The Coordination and Regulation of Movement, Pergamon, Oxford, 1967

25. V. B. Brooks, The Neural Basis of Motor Control, Oxford University Press, New York 1986

26. J. Peillard, "Apraxia and the neurophisiology of motor control," Philos. Trans. R. Soc.Lond. Biol., 298:111-134 1982

27. C. Chez, "Voluntary Movement", in Eds. E. R. Kandel, J. H. Schwartz, T. M. Jessel, Principles of Neural Science, Appleton & Lange, Norwalk, CT, 1997

28. D. Parves, et al, (eds), Neuroscience, Sinauer Associates Inc., Sunderland, MA 1997

29. C. G. Phillips, Movements of the Hand, Liverpool University Press, Liverpool, GB, 1986

30. S. Grossberg

31. A. Meystel, Autonomous Mobile Robots : Vehicles with Cognitive Control, World Scientific, Singapore,1991

32. A. Meystel, "Planning in a hierarchical nested controller for autonomous robots," Proc. IEEE 25th Conf. on Decision and Control, Athens, Greece, 1986

33. M. Mignard, J. Malpeli, "Paths of Information Flow Through Visual Cortex," Science, Vol. 251, 1991, pp. 1249-1251

34. P. Churchland, T. Sejnowski, The Computational Brain, The MIT Press, Cambridge, MA 1992

35. D. C. Van Essen, C. H. Anderson, B. A. Olshausen, "Dynamic Routing Strategies in Sensory, Motor, and Cognitive Processing", in Eds. C. Koch and J. L. Davis, *Large-Scale Neuronal Theories of the Brain*, The MIT Press, Cambridge, MA 1994

Glossary

Architecture—is a network of relations among the entities; a set of objects of interest is called architecture if there is an inner pattern, or a law that can be attributed to this network.

Category, categorization—A unit of classificatory division performed for a multiplicity of objects and/or events is called a category if under some circumstances it can be considered an entity. A class, a cluster—are categories; categorization is the process of constructing this classificatory division.

Commands—is a codeword for the assignment admissible within the particular level of the system; it is the output of the upper level presented in the form of the encoded task. The command by itself is not sufficient to trigger the operation: the State, and the Spatio-Temporal Model should be submitted by the World Model. Commands are usually sent in a form of temporal strings.

Complexity—Any architecture can be characterized by a number of procedures required to analyze (and or to compute) the processes in this architecture. Complexity is a value proportional to the number of these procedures.

Concepts—Concept, or category, may be considered a result of extracting the essence of generalized objects and/or events. Their features are represented by sets of attributes. Groups of generalized objects share attributes. A concept represents a common attribute or meaning extracted from a diverse array of experiences, while a symbol stands for a particular class of objects and/or events. Concepts are used to sort specific experiences and map them into general rules or schemata which represent linkages existing between objects and events.

Connectionism—is a way of analysis by modeling the system as an explicit network for the subsequent computational modeling. Connectionist approach does not require prior generalization upon this network. Connectionist hypotheses of motion presume taking in account a multiplicity of "commands" and/or "actuators" simultaneously (which requires simulation).

Control—is a joint process of generation and execution of the strings of commands to provide functioning of a system. The term "control" is used at higher resolution levels to describe the same phenomenon which is called "planning" at the lower level of resolution.

Cybernetics—is a scientific paradigm which demonstrates that all objects and systems can be represented by using models with feedback (similar to Figure I of this Foreword). This paradigm was proposed in 19th century by Ampere and actively pursued in 20th century by N. Wiener.

Cybernetics of the mth order—One of the N. Wiener's contemporaries, H. von Foerster introduced a concept of "cybernetics of the 2nd order" for modeling of systems with self reflection (whose World Models contain also models of themselves). It became clear that functioning of the system with self-reflection can be more adequate if their model of themselves would include the model of the world supplemented with their model of themselves, and so on. This is how the mth order emerges; the value of m depends on the number of recursive repetitions of this consideration. Obviously, nervous system is a system which allows for multiple self-reflections.

Deliberative motion—This term is used in the area of Cognitive Science to distinguish a class of motions which are contemplated by constructing possible alternatives and subsequent selection of the best of them.

Elementary Loop of Functioning—is a unit of analysis of systems which must include Sensors, Perception, World Model, Control Generator, Actuators, and a relevant part of the World. This unit should be sufficient for discussing each process of interest.

Entity—is something that can be given a name, it is characterized by some unity of all its components.

Feedback control—is a string of commands prescribing a correction to the plan; it is computed by comparing the outcome of the process with the planned trajectory without correcting the world.

Feedforward control—is a string (time sequence) of input commands, and/or input signal determined as a function of time (or another variable assumed independent within the system under consideration) which is supposed to provide the desirable output motion trajectory.

Generalization—Generalization presumes unifying a set of features and/or property and/or object into one property, or object (generalized property, object). There are many methods of generalization including generalization via approximation, via averaging, via integration, via aggregation based on recognition and detection. Generalization always ends with labeling, or relabeling.

Granularity (see also **Scale** and **Resolution**)—is the process of discretizing the system (or its state space) into minimal intervals (undistinguishability units). Each of these units is considered to be half a unit of scale. The inverse of this unit is used to evaluate the resolution of the model.

Incompletely known systems—are system which should be controlled by existing models although it is known that these models do not reflect all factors; these factors are simply unknown.

Inverting the motion—If the desired output motion is known, and the model of the system that produces this output is known, then an input command string can be found. Any procedure of finding it can be called "motion inverse"

Knowledge—is a bulk of all available units of information, properly labeled and organized into a relational network with a particular architecture; it implies our ability to use it efficiently for reasoning and decision-making.

Learning—is a process of acquiring knowledge from experience; it is to be used for modeling the world and our behavior in it.

Level of hierarchy—is a representation of the system with a particular level of detail. Level of a hierarchy can be also called level of resolution (or level of granularity, level of generalization, level of abstraction).

Look-up Table—is a table which contains an exhaustive list of objects, relations, and/or rules of interest.

Nestedness—a property of *being contained in*. In this foreword, we apply this term as related to knowledge and information. This means that nesting can always be demonstrated via interpretation. Nesting puts some conditions upon information processing within hierarchies.

Object-oriented - This adjective is used for an approach which is widespread in the contemporary programming and knowledge organization. The essence of it is the organization of information in a hierarchical (multiresolutional, multigranular, multiscale) manner.

Resolution (see also **Granularity** and **Scale**)—is an inverse of the unit of scale.

Rules and Metarules—are logical statement of cause-effect. Rules assert that a particular antecedent implies a particular consequent at a particular level of granularity. A multiplicity of rules can imply a class of rules. The general statements of cause-effect for the class of rules we call a metarule.

Scale (see also **Granularity** and **Resolution**)—a system of discretization chosen for measuring the variables. Usually, the unit of the scale is twice the indistinguishability zone.

Schema (pl. *schemata*)—is an abstract representation of the distinctive characteristics of associated objects and/or events which contains information about implication of their association. Frequently, schemata are used to describe rules, or implications. They demonstrate what is the effect associated with a particular cause.

Scope of attention—in the subset of the world which is selected as the subset of interest; it is assumed that the rest of the world cannot affect our decisions about the scope of attention.

Semiosis—is the joint process of functioning/learning within an elementary loop of functioning. Semiosis is always associated with producing new signs. The latter emerge as a result of generalization of experiences.

Semiotics—is a science of signs; it is a science of organizing knowledge into a symbolic system. Semiotics is involved into the essence of the processes of intelligence, learning and functioning of living systems. There are many areas of semiotics focused on more narrow domains of knowledge (biosemiotics, zoosemiotics).

Signs—are notations associated both with an interpretation and with a real object and/or event.

Symbols—represent the next level of abstraction from experience; they are arbitrary labels for things, qualities, events, actions, etc. Whereas a sign represents a specific experience, a symbol is a representation of generalized objects and/or events, i.e., a symbol is attached to clusters of similar signs. Symbols are used in information of higher cognitive units called concepts.

Symbol grounding—is a procedure of putting in correspondence thie encoded knowledge and the experimental results required to verify the encoded knowledge.

Alex Meystel
*Drexel University and National Institute
of Standards and Technology*

Foreword

Every thoughtful scientist and clinician trained in America during the past 25 years must at some level be aware of the process by which we as students are *educated*. A consistent theme has been, and continues to be, the acquisition of numerous facts and details for recollection at the time of formal testing. What is reinforced and rewarded is not one's ability to *think* or *solve problems*; the overall trend in the American educational system is to deliver a body of knowledge or information—whether useful or not—in such a manner as to facilitate its recall during those moments in which the rewards for its precise recollection are greatest. It was quite encouraging to me during my years of formal education to be informed primarily by my European-trained colleagues of the existence of educational processes in countries other than my own where emphasis and importance are placed on developing in its students the ability to think and to problem-solve, and not simply to acquire a body of knowledge. Despite the effectiveness of such enriching methods of teaching, thoughtfully and critically calling into question the fundamental assumptions of the biological sciences is rarely rewarded, and reasonable explanations that help to resolve many of its glaring disparities continue to elude even the brightest of its scientists. For example, the dogma of contemporary neuroscience considers language itself to be a result of vague phenomena that occur within the functional architecture of the cerebral cortex. However, clinical neuroscience has demonstrated evidence in humans for the disruption of unique linguistic functions following injuries confined to the basal ganglia—a *subcortical* structure.

Among its many contributions, contemporary neuroscience has provided a wealth of experimental data that reveal intrinsically inseparable relationships between the many varieties of methodological techniques available to brain research and the functional features of brain and mind—both great and small—that they attempt to explain. However, it is the theoretical foundation upon which these explanations are based that restricts further expansion of our current understanding of how the nervous system works and our ability to apply this understanding of how the nervous system works and our ability to apply this understanding and its principles to practical issues such as the management of neurological diseases. While the zeitgeist of contemporary neuroscience supports the generation of enormous amounts of experimental data with little emphasis on how these data affect the fundamental theoretical assumptions of neurobiology, such an approach seems disquietingly similar to the current educational process in America. Many scientists—including myself—are frustrated with the lack of a unified conceptual and theoretical framework in the brain sciences that adequately addresses the countless disparities that arise between empirical findings and their interpretation when the current theoretical paradigm of neurobiology is utilized. It is the inadequacies of the current theoretical paradigm for neurobiology that are addressed by the content of this provocative and timely theoretical work.

Within the pages of this monograph, Dr. Konstantin V. Baev presents to the reader an alternative conceptual theoretical framework for contemporary neuroscience that provides a set of unifying principles of *functional* construction of the nervous system. Using examples that include the involvement of spinal cord mechanisms in automatic behaviors such as locomotion and scratching, as well as the role of the frontal cortex in higher, more abstract functions such as cognition, Dr. Baev reveals how identical functional principles of construction accurately and appropriately describe the fundamental nature of the control of *any given feature of the nervous system*. The fundamental assumption that provides the unifying theoretical basis for his novel heuristics is that, from a conceptual standpoint, *all* behaviors expressed by the nervous system— whether *inborn* or *acquired*—are *automatisms*. Once the psychological barriers to the acceptance of this fundamental notion are overcome, the utility of this alternative theoretical paradigm for nervous system function becomes plausible and understandable. The use of this novel conceptual theoretical paradigm greatly facilitates and explanation of the role of the basal ganglia in language function, and how lesions or stimulation of identical regions of the motor thalamus or ventral pallidum are effective in the surgical management of Parkinson's disease.

The structure of the text itself outlines the conceptual and theoretical transformation of Dr. Baev's own understanding and appreciation of the function of the nervous system. His contributions to systems neurobiology in the context of the control of automatic behavior by *Central Pattern Generators* in the spinal cord are well known and groundbreaking. With the publication of this present work, Dr. Baev has once again provided a challenging, new, and potentially fruitful direction for a generation of neurobiologists who seek to unearth a deeper and more fundamental understanding of the function of the human brain—our "two pound universe." Using concepts from physics, mathematics, control theory, and computational neuroscience and neurocomputing, the author presents a functional perspective with the assumption that biological neural network systems are designed by nature to efficiently and accurately regulate the automatisms of the nervous system. Dr. Baev postulates that the means by which this self-regulatory process occurs is the function of an *optimal control system* for these automatisms that incorporates in its complex computational abilities a capacity for learning. Its features are well-detailed and lucidly presented.

The alternative conceptual theoretical paradigm for neurobiology presented in this monograph holds the promise of substantively and fundamentally advancing neuroscience education for a generation of new students that must not be easily satisfied with mere memorization of facts and details, or the generation of endless amounts of experimental data in the absence of an adequate theoretical framework for its rational and meaningful interpretation. An understanding and appreciation of this small number of universal principles for constructing a functional framework of the nervous system will doubtlessly modify existing pathophysiologic, diagnostic, and management approaches to neurological disease. Application of these conceptual principles to the computer sciences will result in a revolutionary expansion in the computational abilities of network systems developed by neurocomputing and computational neuroscience. The implications of such an expansion for commercial application in the computer and communications industries are potentially staggering. An experience with and understanding of Dr. Baev's novel heuristics inevitably leads one to the conclusion that, whether an individual reader agrees or disagrees with the hypotheses put forth in this monograph, the reader as well as the discipline of neurobiology itself will *never again* view the function of the nervous system in quite the same way.

Karl A. Greene, M.D., Ph.D.
Institute for Neurosciences and Division of Neurosurgery
Conemaugh Memorial Medical Center
Johnstown, Pennsylvania
April 1, 1997

Introduction

It is hard to express the current situation of modern neuroscience better than it was done by Graham Hoyle a little more than a decade ago. In 1984, he wrote: "Unfortunately, in spite of an explosion of research activity in neuroscience in the 34 years since the Cambridge meeting, there has been little advance in its conceptual underpinnings. The single general framework that has ever existed, the McCulloch-Pitts (1943) binomial model of neural function, had to be abandoned when intracellular recording revealed the widespread occurrence and importance of analog information processing and signaling. But the vacuum left behind has yet to be filled with even a tentative new model. Neuroscience came to be the art of the do-able, with expediency ruling the day, rather than a soundly based intellectual domain. Three generations of neuroscientists have now been trained without any link to a widely accepted general theory of neural circuit function and neural integration. They have been given to believe that they are engaged in a massive fact-finding operation guided only by the relative softness of the seams in the body unknown that happened to face their individual picks! Science without larger questions provides a dismal prospect to a truly inquiring mind. Of course, to those who would make careers out of providing random facts, nothing could be nicer, so varied and so complex are nervous systems. There is enough material to occupy armies of such persons for centuries. But, without some strong delineations, neuroscience will continue to explode into myriad fragments. We shall end up with masses of descriptive minutiae of many nervous systems without advancing our overall understanding of how they do the job for which they evolved." (Hoyle 1984, p. 379).

Graham Hoyle was not the only scientist who thought this way. A similar evaluation can be found in the 1982 abstract of a Neurosciences Society Workshop on motor control. Here it is stated that "Experimentation in this area cannot proceed without theories" and "the predominant, almost ex-

Biological Neural Networks
Konstantin V. Baev
© 1998 Birkhäuser Boston

treme, empiricism which characterizes research in motor control might in the long run be disadvantageous for progress" (Ostriker et al 1982, p. 155).

The above statements do not require further comments. We have to accept that almost all scientists think the above situation has not significantly changed since these statements were made, despite the fact that the 1990s were proclaimed by scientists and politicians as *the decade of the brain*, and there are already proposals to announce the whole next century as *the century of the brain*. I believe that the majority of scientists consider such announcements as political moves that have a strong impact on brain sciences only because they help to attract greater financial resources for brain research, without necessarily relating to any real progress in the brain sciences. The abundance of new terms introduced by the brain sciences during the last several decades—terms such as "neuroscience," "neurobiology," and "computational neuroscience," which accentuate the interdisciplinary nature of the brain sciences—partly supports this belief, because it is easier to solicit funding for research by using the sophisticated image that an interdisciplinary program provides. The old term "neurophysiology" has been abused to the extent that it no longer possesses the capacity to attract monetary resources, even though neurophysiology has always been an interdisciplinary science. Nevertheless, this political strategy also reflects a strong desire to finally make a breakthrough in our understanding of the brain by conducting more experimental research, as if to say, "Let us do more intensively what we can, and a new quality in our understanding of the brain will somehow emerge."

One can easily identify the simplest explanation for the lack of a reasonable theoretical framework for the neurosciences. It is the enormous complexity of the nervous system. All theoretical ideas and concepts about brain construction can be subdivided into two groups—biogenic and technogenic. The biogenic philosophy is rather popular in contemporary neurobiology because the application of simple technical concepts for describing brain function usually fails. Classical textbooks of neurobiology and the majority of modern theoretical and experimental scientific research are mainly based on the biogenic concept. This philosophical approach denies the usefulness of the technogenic approach to neurobiology, and accordingly maintains that any fruitful theory of brain function can only be created by exclusively using and developing biogenic concepts and ideas. This point of view is quite strange because almost all theoretical and experimental achievements in neurobiology are based on discoveries from other scientific disciplines. For example, the origins of the reflex theory and the program control principle, as well as any experimental method used in neurobiology, can be found in one or several different scientific disciplines—chemistry, physics, and mathematics, for example. The biogenic point of view is also not constructive because it ultimately denies the usefulness of the knowledge accumulated by numerous disciplines related to neurobiology. Let us remember, for example, the theory of vitalism that was very popular in the nineteenth century. The achievements of molecular biology in the twentieth century proved conclusively that it is not necessary to propose that life processes arise from some nonmaterial vital principle and cannot be explained entirely as physical and chemical phenomena.

Biological neural networks are created by nature, and the laws of nature should be applicable to them. The fact that the application of simple technical ideas to describe brain function usually fails (a fact used by those who support biogenic ideas) has a clear explanation: brain complexity surpasses the complexity of simple physical or other systems for whose behavior descriptive laws are already known, and it is obvious that the description of a behavior of a complex system cannot be reduced to the description of a behavior of a much simpler system. Clearly, an investigation of the brain, one of the most sophisticated products of nature, will need the joint efforts of numerous scientific disciplines and new theoretical approaches.

One of the best ways to view the brain is as a control device created by nature, and the goal of neurobiology is to understand how it works. There is also another field that deals with control problems, but does so by creating artificial controlling devices. Recently, this field has expanded so broadly that it is hard to find a science or a technological field where artificial controlling devices are not used. Control theory, or cybernetics, and its derivatives—computer science, neurocomputing, and artificial intelligence, among others—belong to those sciences that are attempting to create such artificial control devices.

Obviously, there should be common interests, theoretical frameworks, and methods of research in neurobiology and control theory. However, these sciences have developed independently as separate sciences that share very little common ground because they initially had different goals: to analyze the existing controlling device in the first case, and to create a controlling device capable of solving a definite control task in the second case. This difference in goals predetermined the research strategies that have been used in neurobiology and control theory, strategies of analysis and synthesis, respectively.

In any science, analytical and synthetic approaches are complementary. Both depend on the existing theoretical framework, and there should be an optimal relationship between them in order for that science to survive and succeed. If one approach prevails, the results of such an asymmetry can be costly, time-consuming, and deleterious for any science. This simple truth is intuitively well understood, but experience shows that their relationship is often far from being optimal. At each stage of the development of any science, one of these approaches prevails, and quite often the prevailing approach can reach an absurd level. This is exactly what has happened in neurobiology, in which the analytical approach has been the dominant one for many years, and the synthetic approach has usually been conducted in a very specific mechanistic manner (see below).

Neurobiology has developed as a descriptive experimental science, while control theory has always been a highly technical discipline. Each of these sciences has developed its own language and theoretical basis. As a rule, our theoretical views reflect the level of our understanding of cause-effect relationships and determine the way we study them. One must have at least a working hypothesis in order to conduct any kind of experiments. A bad theory is better than no theory, because it can lead to testable experiments whose results may contradict the original theory, thus initiating the creation of a new theory. A good theory usually possesses a highly predictive, and hence constructive, power, and research significantly benefits from it. However, any theory has its limitations in explaining the whole variety of experimental phenomena. An understanding of the nature of those limitations usually helps to create a new theory or to modify an existing one.

The more knowledge accumulated by both sciences, the stronger the basis created for their mutually beneficial interaction. Presently, this interaction has started to emerge, and we are witnessing significant changes in the theoretical bases of both sciences. One of the best examples of this is the appearance of new sciences such as neurocomputing and computational neuroscience which use similar approaches to analyze the function of artificial and biological neural networks, respectively. There is also no doubt that such beneficial interaction will inevitably lead to the creation of a new brain theory, probably a unifying brain theory, and to new artificial control devices that will be based on the new principles of brain function.

Let us try to make an analysis of the theoretical bases of both neurobiology and control theory, at the most general level, in order to understand what new ideas can be adopted by neurobiology from control theory, and vice versa. A discussion of the fundamental concepts that have played a significant role in determining the development of both sciences throughout much of their history would be the best starting point for such an analysis. Such concepts include reflex and program control in neurobiology, and the concept of optimal control in control theory. In neurobiology, the abovementioned ideas of reflex and program control are considered classic and are the theoretical basis of numerous textbooks of neurobiology (Shepherd 1994; Kandel et al 1991) and neurology (Adams and Victor 1993), and of the majority of experimental and theoretical neuroscience research. Gordon Shepherd's abovementioned textbook of neurobiology is one of the best, and is translated into many languages. However, already in the preface to his textbook, Gordon Shepherd mentions the necessity of creating new neurobiological concepts, because experimental facts are generated by the contemporary neurosciences much faster than concepts capable of explaining them. It is also necessary to mention that even in computational neuroscience—presently considered by many to be the last word in neurobiology—classical views are still dominating.

We shall thus start from an analysis of the most popular, and hence most influential, theoretical concepts in neurobiology. To analyze all such ideas that can be found in neurobiology would be inappropriate for our purpose, because such a discourse could appear clumsy and pointless. In modern neurobiology, which has developed primarily as an experimental descriptive science, researchers use so many new terms that theoretical conclusions may be difficult to reach, because it is hard to carry on a systematic discussion of the necessary ideas. A lot of those terms, either created within neurobiology or accepted from other sciences, are used with a considerable degree of vagueness. They are quite often introduced without strict definition, or accepted from other sciences without a clear and appropriate understanding. As will be seen below, "central pattern generator" is one of the vague terms that is still broadly used in neurobiology. Unfortunately, such terminology is an undeniable component of the language commonly used by modern neurobiology which is why some of these terms will be used in the beginning of the discourse to follow. Their inappropriate use will be explained in light of the concepts being analyzed. We shall also analyze several theoretical ideas that have been proposed in an effort to complete the reflex theory and explain the experimental facts that were not accounted for by the original reflex theory.

Classical views teach us that the nervous system has to be understood in a mechanistic perspective; that is, biological neural networks must be understood in terms of interactive nerve cells, transmitters, and membrane properties. Experimental analysis is directed toward obtaining this type of information, while a synthesis is understood as a "recreation" of the system, and the model thus recreated has to mimic the behavior of the studied system as precisely as possible. This combination of approach defines how the relationship between structure and function is viewed within the mechanistic approach. As we shall see below, the mechanistic approach has very serious limitations. Within the mechanistic framework, both analysis and synthesis appear to be "rather effective" only for analysis of simple invertebrate and vertebrate networks. This approach assumes that the network being analyzed is unifunctional. Mammalian networks, which are as a rule multifunctional, have proven to be too complex for such an approach. The major reason that contemporary research utilizing simple networks is so popular is that it eventually leads to an approach of "let us study what we can instead of studying what we should." The tacit hope in such a scenario is that perhaps the knowledge of simple systems will help us to understand the function of complex systems. Historically, science has almost never worked this way. It has been demonstrated countless numbers of times that studying a simple object usually only results in simple knowledge. Such a mechanistic approach does not possess much power to generalize results, if any.

There is another drawback to this mechanistic approach. It generally considers any neural network as a prewired entity, and thus there is not enough flexibility within this approach to analyze the process of self-organization.

At the most general level, the foregoing statement implies that the theoretical basis for contemporary neurobiology does not possess the necessary functional completeness to be capable of describing the behavior of complex biological neural networks. At most, within the framework of the mechanistic approach it is possible to formulate such a problem as how a neural circuit is built, but not why it is built this way. The following obvious question appears: should someone try to complete the existing theory, or should it simply be abandoned and replaced by a new one? The second alternative is the avenue advocated in the present book. This new theory will require the implementation of several additional notions such as automatism, the network computational principle, optimality, and some other related concepts. Clearly, a new theory has to be capable of explaining all the experimental facts accumulated to the present day, possess a high predictive power, and bring a new understanding of the relationship between structure and function for complex multifunctional networks.

The accumulated experimental facts demonstrate the existence of an enormous variety of neural network schematic designs (including synaptic, cellular, and molecular mechanisms) in different animal species and in different parts of the nervous system within the same species. This variability is commensurable with the variety of life forms having a nervous system. The new theory helps to reveal common principles of functional neural network construction, principles that cannot be revealed by using a classical neurobiological theoretical framework. The existing mechanistic approach limits scientists to studying the concrete constructions of all these networks. This new theoretical approach is based on the fact that the capacity for automatic control, inborn or acquired, is the very essence of the nervous system. Therefore, if we understand the principle of construction of a given neural automatism, then we will understand the function of the brain. The nervous system inherited control principles from those living control systems that existed in nature long before the appearance of the brain. Thus it is also obvious that the notion of an automatism can be applied to other biological control systems that do not have neuronal organization but possess features of functional organization. This will eventually result in a very broad applicability of the brain theory to other biological disciplines.

Numerous experimental facts and several quite popular neurobiological problems are analyzed in this book. This is done with one major reason in mind: to demonstrate the applicability of the theoretical concepts developed in the pages to follow. However, it is not the goal of this book to present a systematic description of *all* the functions of the nervous system. The hope is that the reader will be able to easily apply new theoretical concepts to any subsystem of the brain after reading through this book.

1

Limitations of Analytical Mechanistic Approaches to Biological Neural Networks

The primary theoretical principle—reflexion or reflex—that has determined the development of neurophysiology, and still continues to do so, is about three hundred and fifty years old. This classical point of view teaches us to understand the nervous system as a device that functions according to the formula "stimulus-reaction." The theory goes back to the sixteenth century, when Rene Descartes developed the first scientific reflex theory, based on the knowledge of that time regarding the motion of liquid through a system of connected vessels that possesses a pump and valves. In fact, it was the first technogenic idea that was used to explain brain function.

The basic formula "stimulus-reaction" was preserved in later reflex theories. In the beginning of our century, neural centers were considered to be analogs of telephone stations. Obviously, this idea did not change the principle of the reflex theory, although its technical point of view became electrical instead of mechanical. One can consider the reflex theory as the simplest mechanistic means of explaining cause-effect relationships found in the central nervous system in terms of specific routes that can be inborn or acquired (learned). According to the theory, signals evoked by a stimulus are delivered via corresponding reflex arcs to an effector organ, and this mechanism is considered to embody the very essence of nervous system function.

The reflex theory was easily accepted because of its conceptual simplicity. According to this theory, the understanding of brain function requires a knowledge of the numerous reflexes inherent in a given species of animal. Thus, the behavior of a particular animal can be understood if the structure of reflex arcs and the signals transmitted via corresponding arcs during a concrete behavior are known, and the interaction of the numerous corresponding reflex arcs are balanced by using excitatory or inhibitory influences on neurons that are included in the reflex arcs under study. The efforts of many generations of physiologists were focused on obtaining such infor-

Biological Neural Networks
Konstantin V. Baev
© 1998 Birkhäuser Boston

mation. With the appearance of more powerful and more precise methods of neuronal activity analysis, scientists aspired to obtain more detailed data on reflex arc construction and mechanisms of reflex interaction. Within the reflex framework, learning was considered to be the establishment of new reflex arcs, and the problem of learning therefore appeared to be rather simple: one had only to investigate the mechanisms by which a new reflex arc is established, and if a set of learning rules is established it will eventually lead to the solution of the problem of learning. Hebbian learning rules (see Section 1.2) are a typical example of the simplistic perspective utilized in the problem of learning within this reflexive theoretical framework.

Obviously, the reflex theory directed researchers to investigate how the brain is constructed. On a larger scale, this approach provided some degree of scientific direction and advancement, and we have to accept that the majority of the experimental facts that were accumulated by neurobiology were obtained through research based on reflex theory. However, as we shall see below, this theory had only a limited power to formulate, and consequently to answer, the following question: why is a brain created the way it is?

It is easy to find supporters of the belief in either discrete or universal applicability of the reflex concept. Thus, C. Sherrington (1947) belonged to the group of supporters of the limited applicability of reflex theory for describing nervous system function. He proposed that only the functioning of the lowest neuronal levels could be described using reflex theory. Instincts, acquired habits, and spontaneous activity are not reflexes, according to Sherrington. The belief in a more universal applicability of the reflex concept can be found in the works of I. Sechenov and I. Pavlov. In the last century, the Russian physiologist I. Sechenov tried to describe the work of the cerebral cortex as a reflex in his book *Brain Reflexes* (1863). He considered thoughts to be reflexes that were being delayed. According to Pavlov (1949), who introduced the term "conditioned reflex" into physiology, " . . . the enormous part of higher nervous activity is largely, if not wholly, explained by the physiologist on the basis of conditioned reflex."

The limited concept of reflex suggested initially that the reflex principle of brain functioning needed something more, and the universal concept postulated that the work of the brain could be investigated and completely explained on the basis of reflex theory. Subsequent developments in neurophysiology, however, proved to supporters of the universal reflex concept that the reflex theory was inadequate to explain a whole variety of nervous system functions.

Reflex theory was then confronted with new stumbling blocks. Only simple reflexes appeared to be explainable using the reflex approach. In the case of complex reflexes involving numerous levels within the nervous system, the detailed investigation of arcs appeared to be embarrassing, if not practically ruled out. It was not even clear how to explain the existence of parallel chains in the same reflex arc from the perspective of reflex theory.

Tracing cause-effect relationships turned out to be impossible for complex behaviors that were initially explained as the temporal interactions of numerous reflexes. For example, within the limits of the idea of a reflex ring, the efferent impulses are not considered as the final result of reflex action, but rather as a source of new excitatory influences on the neural center. Some complex behavioral reactions are significantly delayed in response to stimulus, and some have appeared without any obvious reason. In such circumstances, the presence of memory traces or some other processes in the nervous system has been proposed, such that a neural center was enriched with its own dynamics. The notion of the dynamics of a neural center was thus outside the limits of the reflex principle. However, this notion only enables one to somehow explain complex behavioral events with regard to the variability of reflex reactions in intact animals and explains why they differ from those stereotyped reactions observed by neurophysiologists in model experiments (in animals under general anesthesia or after lesion of higher brain levels).

Numerous investigations have shown that self-dynamic ability is a fundamental property of neural centers. Obviously, given factors explaining intrinsic regularities that characterize neural center dynamics, it becomes possible to obtain the answer to the question: how does the brain work?

The first attempts to complete the reflex theory were made at the end of the first half of the twentieth century and were directed at explaining complex behavior. The idea of program control was proposed (Bernstein 1966, 1967, 1984; Lorenz 1950; von Holst 1954). The program control principle was, as a matter of fact, the first attempt to explain why a neural center has its own dynamics. The notion of a *fixed action pattern*, which is a cornerstone of the ethology movement, is in fact a generalization of the program control idea. This term defines complex stereotyped acts (programs), including specific temporal sequences of actions inherent to a specific species of animal. External or internal stimuli play the role of a trigger that gives rise to a fixed action pattern. But these notions did not explain the principle of program realization in neural centers.

Obviously, the idea of program control does not belong to reflex theory and was not created within it. This idea has a technogenic origin and was easily adopted from control theory. The basic notions of control theory will be discussed in Section 2.2. The term was mentioned here only because it is broadly used in neurobiology. The way the concept of program control influenced neurobiology is ideologically very similar to the way the reflex theory determined the direction of experimental neurobiology research. Experimenters were trying to understand the neural circuitries that were responsible for the generation of different programs, i.e., to study how those circuitries are constructed. Very simple logic motivates this approach: each program is generated by a specific circuitry, and the goal of science is to discover the cellular, synaptic, and circuitry mechanisms involved in the realization of the program.

The reflex concept underwent a certain transformation with the appearance of the program control principle. It was also enriched with dynamics, and scientists began to consider the reflex response as dependent on the state of the neural center involved. The postulation of the existence of correction reflexes was put forward. According to this proposal, the program generated by the center adapts to external conditions by using various reflexes.

To make our discussion more concrete, let us analyze the limitations of the aforementioned classical theoretical framework of neurobiology by using as an example the control of inborn automatic behaviors and of learning, in animals, according to a classical conditioning paradigm, Pavlovian association learning. The majority of research in these fields was based on classical theoretical views, and let us accept temporarily that we also hold these views.

1.1 Inborn Automatic Behaviors From the Point of View of Classical Theory

Inborn automatic movements, such as locomotion, breathing, swallowing, chewing, and scratching, are the most extensively studied of all inborn behaviors. Inborn automatic movements can be generated by corresponding neural centers without the aid of peripheral afferent feedback, which is why those centers were referred to as "central pattern generators" (CPG). The illustrative example of the locomotor generator will be used to familiarize the reader with the common contemporary strategy of CPG investigation.

According to existing traditional views, the generators produce motor patterns for different body parts—limb, wing, etc. The interaction of different generators is the basis for the coordination of different body parts. Generator activity is under the control of peripheral feedback and higher motor centers. The initiation of the CPG's output is accomplished by simple descending or afferent commands that activate it for a specific behavior (Figure 1).

1.1.1 A Brief Review of the Locomotor Behavior Evolution in Animals

Let us briefly consider the evolution of locomotor behavior in animals, in order to imagine its enormous variety in the animal world. In the course of evolution, animals adapted to different surroundings through the evolution of the species themselves and through the simultaneous development of their movements. Moreover, evolution of the overall body form, its different

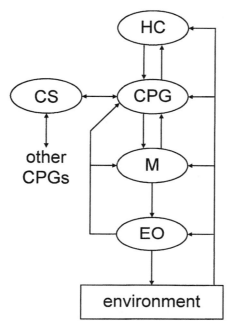

Figure 1—Current views of the system controlling locomotion. HC, highest centers; CPG, central pattern generator for a given body part; M, motoneurons; EO, effector organ; CS, coordinating system.

structures, and their functions was tightly coupled with the appearance of specific forms of motor behavior.

The simplest forms of locomotion are observed even in single-celled animals. They move with the help of pseudopodia, cilia, and flagella. These forms of locomotion are preserved in the simplest of multicellular organisms and in certain cells of higher animals.

Special anatomical structures, such as muscles, the skeleton, and special appendages, appeared in the evolution of multicellular organisms. A skeleton allows for the effective application of forces generated by muscles, which significantly improves movement. A hydrostatic skeleton is found in invertebrates. The primary forms of locomotion in animals with a hydrostatic skeleton are slow locomotions such as swimming and crawling. External and internal skeletons, stiffer than their hydrostatic counterparts and hence more effective, are found in arthropods and vertebrates, respectively. The appearance of a stiff skeleton was a real revolution in the evolutionary development of movement. This type of skeleton, in particular its extremities and their lever function, appears to be very effective during movement along a firm substratum, and especially effective for flight. Despite the variations in overall body structure between vertebrates and invertebrates, their principal modes of locomotion are similar—swimming, crawling, walking, running, and flying. Vertebrates have less variability in overall body structure and are less numerous than arthropods, but can compete with them in accommodation to different surroundings and actually surpass arthropods in their ability to react to external irritants.

It is possible to draw several conclusions when considering the evolution of form and structure of an animal body from the point of view of locomotor function. In the course of evolution, decrease in the degree of body symmetry (bilateral symmetry in relation to the longitudinal axis) appeared, and corporeal segmentation took place. The appearance of a longitudinal axis resulted in movement of the animal in one of two possible directions, and rostral orientation became more effective for navigating within the environment. Because the rostral end of the body was the first to encounter prey or danger, it was supplied with better receptors that included distant ones (antennae, whiskers, and vibrissae, olfactory, visual, and acoustic sense organs, etc.). This evolutionary process provided the basis for the initial appearance and subsequent improvement of the brain. This organ began to determine the goal of locomotion and utilized this goal to initiate and coordinate the movement of other body parts. Bilateral symmetry and segmentation significantly simplified the solution to the problem of movement coordination by the nervous system, and left their imprint on the construction of the nervous system. A pair of ganglia or bilaterally symmetrical portions of the neural tube (spinal cord) corresponds to each body segment. The degree of complexity of the brain and segmental nervous system is described by a simple rule: the simpler the locomotor behavior of the animal, the simpler is its nervous system.

1.1.2 Initiation of Inborn Automatic Behaviors

Two main ways of initiation of motor automatic reactions, phasic or rhythmic stereotyped movements, are well known: afferent and descending.

Afferent activation

After disconnection of the highest centers from the lowest ones, different types of afferent inputs (tactile, painful, etc.) may initiate the same type of motor activity, even within a single animal species. It can be withdrawal reaction, escape reaction, or locomotion. That is why such an activating influence of afferents is frequently classified as nonspecific. Most frequently, the term "afferent activation" is used in the locomotor paradigm. The neuronal networks controlling these motor reactions in the simplest animals may be very simple, and in the case of a withdrawal reaction may include one synaptic transmission.

In spinalized lampreys, sharks, some types of bony fishes, and tadpoles, stimulation of caudal parts of the body evokes an escape reaction and locomotion (Grillner 1976; McClellan and Grillner 1983, 1984; McClellan and Farel 1985; McClellan 1986). During the escape reaction S-shape flexion of the body takes place, as a result of which the tail part of the body moves away from the irritant. Escape reactions evoked by caudal stimuli do not differ significantly between intact and spinalized lamprey. But the locomotion that follows the escape reaction is unstable and short in spinalized animals, in comparison with the intact ones.

The flexor reflex is a typical example of a withdrawal reaction in the higher vertebrates, cats for example. A detailed description and numerous reference to literature sources may be found in a comparatively recently published review written by Schomburg (1990). The receptive field of this reflex is extremely wide. It is common to unite all the afferent sources of this reflex by the term *flexor reflex afferents* (FRA). FRA include cutaneous and high-threshold muscle afferents. In natural conditions this reflex may be easily evoked, not only by painful stimuli but also by tactile stimuli. The reflex arc of this reflex has not been described in detail, despite numerous investigations devoted to this question.

In cats, in the acute period after spinalization, strongly painful stimulation of the perianal region may evoke unstable locomotion (Sherrington 1910a). Subsequent investigations showed that activation of spinal locomotor centers may be achieved by stimulation of peripheral nerves, dorsal roots, or dorsal funiculi (Budakova 1971; Viala et al 1974; Grillner and Zangger 1974; Edgerton et al 1976). Thin high-threshold afferents exert the most powerful activating influence on locomotor generators.

The efficacy of nonspecific activation differs in different animal species, and depends on the degree of the development of the highest centers. After decapitation, stimulation of afferents easily evokes locomotion in invertebrates and lower vertebrates. In these animals, it may even appear spontaneously (e.g., in the sea angel). In the dogfish, the spinal cord disconnected from the brain generates unceasing swimming movements (Lissman 1946; Grillner 1974). Subsequent development along the vertebrate evolutionary ladder is accompanied by the decrease in the ability of nonspecific afferents to evoke locomotion. In primates and man, in the case of disconnection between the spinal cord and the brain, nonspecific activation of locomotion is practically impossible.

Descending activation

It is well known that information from distant receptors—visual, acoustic, olfactory—may initiate locomotor behavior. Let us remember the behavior of insects. In most of them, fast transition from light to shadow (usually signaling an approaching enemy) evokes an escape reaction and locomotion.

The notion of so-called command neurons appeared in the course of the investigation of systems initiating locomotion in animals. The term was first used in the description of motor systems in crustacea (Wiersma 1938; Wiersma and Ikeda 1964). It was shown in these and subsequent works that stereotyped motor reactions of varying complexity may be observed in response to stimulation of axons (command fibers) isolated from circummesophageal connectives. Moreover, these reactions appeared either in response to a single shock or in response to tonic low-frequency stimulation.

The range of motor reactions evoked by stimulation of command neurons in crustacea is extremely wide: escape reaction (Schrameck 1970; Wine and Krasne 1972; Bowerman and Larimer 1974b), opening or closing of the claw (Smith 1974), movements of swimmerets, movement of abdominal swimming legs (Hughes and Wiersma 1960; Wiersma and Ikeda 1964; Atwood and Wiersma 1967; Stein 1971; Davis and Kennedy 1972a–c), swimming (Wine and Krasne 1972; Bowerman and Larimer 1974b), defense reactions and positioning of different extremities (Atwood and Wiersma 1967; Bowerman and Larimer 1974a), positioning of the body, turns to the right or to the left, cessation, freezing of movement (Bowerman and Larimer 1974a), and different types of forward and backward walking (Atwood and Wiersma 1967; Bowerman and Larimer 1974b), among others.

As a rule, at one level of the central nervous system, it is possible to find several command interneurons evoking similar motor reactions. For instance, about ten command fibers controlling forward and backward walking were found in crayfish circumesophageal connectives. Each type of motor behavior evoked by a single command neuron is specific, and this specificity was revealed at the level of activity pattern of motoneurons and corresponding motor reactions. Command neurons are not sensory neurons, and they are usually related to interneurons. They receive inputs from different sensory modalities. According to some estimates, the proportion of command neurons in the crayfish nerve chain is 1%.

Detailed data about the organization of neural networks that include command neurons in crustacea are absent. There are also no data concerning their activity during the natural behavior of these animals. Detailed data concerning these questions relate only to giant command neurons responsible for some types of escape reaction. These giant fibers—medial and lateral—terminate directly on motoneurons of tail segments (Wine and Krasne 1972). The medial giant neuron is excited under the influence of visual and tactile stimuli applied to the head part of the body. Caudally applied tactile stimuli excite the lateral giant interneuron. However, the escape reaction is not elicited only by these two giant interneurons. Other types of escape reactions are controlled not by giant but by smaller command interneurons.

The concept of command systems was later supported by numerous investigations on invertebrates (see Kupferman and Weiss 1978). For instance, there are quite detailed investigations of the escape reaction in earthworm, leech and octopus. The giant axon is also the main link in the nerve pathway controlling this reaction. In the leech, locomotion can be evoked by stimulation of only one command neuron (Weeks and Kristan 1978).

In vertebrates, the escape reaction is also expressed in fast change of body position (the startle reaction, which is specific to each animal species) and appears in response to unexpected visual, acoustic, or tactile stimuli. The escape reaction has been most extensively investigated in some species of bony fishes (see Eaton and Hackett 1984; McClellan 1986). During this reaction, the fish's body becomes C-shaped (if the body length is small) or S-shaped (if the body length is large), with the result that the head moves away from the irritant. These movements are stereotyped and may be predicted with a high degree of accuracy.

In lower vertebrates, in most cases the escape reaction may be initiated by a single action potential appearing in one of the two Mauthner cells (M-cell), reticular neurons located at the bottom of fourth ventricle. Mauthner cells receive inputs from different sensory modalities: from the VIIIth cranial nerve bringing auditory and vestibular information, and from impulses from lateral line receptors. The action potential in the M-cell is conveyed via a descending contralateral axon to caudal spinal segments. In small fishes (e.g., goldfish), the conductance time is only 1 ms. This impulse evokes direct and indirect excitation of the motoneurons of contralateral muscles (Faber and Korn 1978). Moreover, interneurons responsible for inhibition of contralateral (in relation to excited motoneurons) motoneurons are also excited.

But the escape reaction may also appear without involving Mauthner cells (see Eaton and Hackett 1984). After lesion of the M-cell on one side, the escape reaction evoked from this side is very similar to the reaction evoked from the intact side. Therefore, in addition to Mauthner cells, there are other neural networks in the fish brain stem responsible for the escape reaction. These other networks have not yet been investigated. Such duplication of neural networks responsible for the startle reaction resembles the previously described duplication of networks controlling the escape reaction in invertebrates.

In contrast to lower vertebrates, in which this reaction has a tendency to be of the "all or none" type, mammals have a startle response that depends on the strength of the stimulation (see Davis 1984). The startle reaction is evoked most effectively by an unexpected strong acoustic stimulus. During this reaction in rats, generalized flexion leading to shortening of the body is observed. The posteroventral cochlear nucleus, the dorsal and ventral nuclei of the lateral lemniscus, the caudal reticular pontine nucleus, and the reticulospinal tract are the main structures taking part in the organization of the startle response evoked by an acoustic stimulus.

Locomotor activity may follow the escape reaction in both invertebrates and vertebrates. But it also may appear spontaneously or as a consequence of afferent activation of corresponding command systems. Locomotor activity in vertebrates may also be activated by tonic (20–50 Hz) electric stimulation of specific brain stem structures. Regions in which electric stimulation evokes locomotion are called locomotor regions (LR). They were found in all investigated vertebrate species: fishes (Kashin et al 1975, 1981; Leonard et al 1979; McClellan and Grillner 1984; McClellan 1986), reptiles (Kazennikov et al 1980), birds (Steeves and Weinstein 1984), rats (Skinner and Garcia-Rill 1985), cats (Shik et al 1966a, b; Mori et al 1977), and monkeys (Eidelberg et al 1981), and they have a number of similar features. Locomotor regions are conjugate in relation to the midline and are found at different levels of the brain stem. They have restricted size; movement of the electrode tip beyond the limits of the region leads to disappearance of locomotor activity. During stimulation of a locomotor region, an increase in stimulating strength is accompanied by more intensive locomotion. Experiments on different animal species have shown that lateral and ventrolateral regions of the spinal cord are necessary to conduct descending commands to spinal generators (Kazennikov et al 1980; Eidelberg et al 1981; Steeves and Jordan 1980; Williams et al 1984).

The most intensive studies of locomotor regions have been conducted on cats. Three main locomotor regions were described: hypothalamic (HLR), mesencephalic (MLR), and locomotor strip (LS). What determines the functional role of locomotor regions? There are a lot of experimental data concerning this question, but a full analysis of them is hardly expedient here. We discuss below only the main conclusions that can be drawn from these data. Both neurons located in locomotor regions and fibers ending in or passing through these regions may be excited during electrical stimulation. Both of these possibilities should be taken into account while making an analysis of the results of different investigations.

The HLR and the MLR are connected with each other and with numerous other brain structures, but the most pronounced efferent projections go to the medial reticular formation (Garcia-Rill et al 1983; Garcia-Rill and Skinner 1986; Jordan 1986; Baev et al 1988). It has been shown that stimulating functionally identified locomotor regions causes them to exert a monosynaptic excitatory influence on most reticulospinal neurons (Orlovsky 1970). During locomotion, either spontaneous or evoked by electrical stimulation of locomotor regions, some of the reticulospinal neurons are tonically active. That is why it is common to explain the effects of stimulation of these regions by activation of reticulospinal neurons, the axons of which pass in the ventrolateral funiculus of the spinal cord. Results obtained in experiments with microinjections in locomotor regions of different pharmacological substances, which influence nerve cells and do not influence nerve fibers, show the important role of neurons located within these regions.

Microinjection of GABA antagonists—bicuculline and picrotoxin—in the MLR led to the appearance of locomotor movements (Garcia-Rill et al 1985). Locomotion could also be evoked by microinjection of glutamate in the medial reticular formation (Noga et al 1988).

Sensory projections to neurons of locomotor regions deserve special attention. The role of these projections was revealed by stimulating corresponding fibers beyond the limits of locomotor regions or by stimulating nuclei—the sources of ascending tracts (Berezovskii and Baev 1988). It was shown that locomotion could be evoked by electric stimulation of the cochlear nuclei, the ventral and dorsal spinocerebellar tracts, the decussation of the medial lemniscus, and the nucleus cuneatus. In the last case, to exclude the spreading of excitation in the caudal direction, the dorsal funiculi were cut below the place of stimulation. Thus, fibers of the medial and lateral lemnisci, which send numerous collaterals to the reticular formation, could be activated during stimulation of the HLR and the MLR.

As for the LS, investigations made by different groups of scientists have confirmed that this strip may be considered a spinal tract of the trigeminal nerve and its nucleus. Microstimulation of the spinal nucleus of the trigeminal nerve in cats (Baev et al 1988; Berezovskii and Baev 1988) or microinjection of glutamate, picrotoxin, or substance P in this nucleus (Noga et al 1988) evokes locomotion. Locomotion may also be evoked by stimulation of the receptive field of the trigeminal nucleus (Aoki and Mori 1981; Noga et al 1988) or by electric stimulation of its branches, the infraorbital nerve for instance (Berezovskii and Baev 1988).

A similar principle of organization may be found upon analysis of the system activating the spinal scratching generator. In the cat, the spinal scratching generator may be activated by electrical stimulation of dorsal parts of the dorsal horn in C_1-C_2 segments (Sherrington 1910b), by mechanical irritation of them (Huang et al 1970) or by the application of chemical substances—bicuculline, strychnine, tubocurarine, or bromphenol blue—to the dorsal surface of these segments (Domer and Feldberg 1960; Feldberg and Fleischauer 1960; Panchin and Scrima 1978; Baev and Zavadskaya 1981; Baev 1991). The effect of the application of these substances is connected with their diffusion in the dorsal horn. In cats, some of the long propriospinal interneurons projecting to lumbosacral spinal cord are tonically active when the pinna is mechanically stimulated.

One more group of experimental facts testifies to the tonic nature of descending commands activating generators for rhythmic movements. Locomotor rhythm generated by isolated pedal ganglia of the sea angel increases under the influence of serotonin (Arshavsky et al 1985a). In vertebrates, locomotor rhythm generated by isolated spinal cord increases under the influence of NMDA, glutamate, or aspartate, and is decreased by NMDA antagonists (Cohen and Wallen 1980; Brodin and Grillner 1985; Brodin et al 1985; Dale and Roberts 1984; O'Donovan 1987; Kudo and Yamada 1987). It is interesting to note that DOPA, a precursor of noradrenaline synthesis, also activates the spinal locomotor generator in vertebrates. It was believed for a long time that the action of this substance was connected with the release of noradrenaline from the terminals of monoaminergic descending neurons (Anden et al 1964), but research on the isolated spinal cord and isolated spinal neurons has shown that the mechanism of DOPA action is different (Chub 1991; Rusin et al 1989; Rusin and Baev 1990). DOPA appeared to be a weak glycine agonist. First, therefore, it increases the efficacy of transmission through glutamatergic excitatory synapses. It has been shown that at low concentrations glycine potentiates responses evoked by activation of NMDA receptors (Johnson and Ascher 1987). Second, low concentrations of DOPA desensitize glycine receptors without evoking inhibitory effects, thus disinhibiting the neural generator network.

1.1.3 The Problem of Organization of Central Pattern Generators for Inborn Automatic Behaviors

Many neuroscientists hope to bridge the gap between our knowledge of neural processes and our knowledge of observed behavior by elucidating the CPG organization. Controlled rhythmic movements are the experimental basis for the current strategy of CPG investigation. As was seen in the previous section, different ways of initiating them in acute experiments have been elaborated. Movements of a body part or efferent commands coming from generator to muscles are used for evaluation of the generator's activity. The removal of motion, which significantly widens the experimenter's abilities, can be obtained in different ways: by sectioning of motor nerves, blocking of neuromuscular transmission (fictitious movement), or by performing the experiments *in vitro*.

At first, we will consider simple invertebrate systems, where the current strategy of CPG investigation has demonstrated its effectiveness, and then the application of this strategy to more complex vertebrate systems will be analyzed. A mechanistic approach prevails in the investigation of simple systems, conducted mainly by analysis. Quite simple synthesis of the results obtained is carried out at the final stage of this approach. According to Getting (1986), "the strategy is basically a 'top-down' approach starting with behavior and working down to a progressively more and more detailed description at the cellular and mechanistic levels." Getting distinguished eight basic steps in the strategy:

1. Description of behavior.
2. Characterization of motor pattern, i.e. efferent commands generated by CPG.

3. Identification of motor neurons and interneurons. Motoneurons are considered noncomponents of CPG.
4. Identification of pattern generating neurons. Any change in the activity of a generator neuron has to be accompanied by changes in rhythm pattern at all levels: interneuronal, motoneuronal, and behavioral.
5. Mapping of synaptic connectivity between neurons.
6. Characterization of cellular properties.
7. Manipulation of network, synaptic, or cellular properties. The goal of this step is to identify the role of individual cells, synapses, or cellular properties in the generation of the overall motor pattern.
8. Reconstruction of the pattern generator, motor output, and behavior. At this final step, the description may be qualitative and verbal, or may be quantitative, when computer simulations are used.

Invertebrate generators

The abovementioned strategy enabled researchers to describe in detail the functioning of a few simple invertebrate generator systems. The classical examples, description of a sea slug's (*Tritonia diomedea*) and sea angel's (*Clione limacina*) locomotor generators (Getting et al 1980; Getting 1981, 1983a, b; Getting and Dekin 1985; Arshavsky et al 1985a–e, 1986b), are presented below.

Tritonia diomedea

In this marine mollusk, swimming can be initiated by transient epithelial contact with the tube feet of predatory sea stars. The response includes an initial reflexive withdrawal followed by a series of 2–20 alternating dorsal and ventral flexions. Swimming movements are correlated with an alternating series of action potentials in two pools of motoneurons.

Twelve identified interneurons, six on each side, form the swimming generator. Moreover, six generator interneurons on each side can produce swimming rhythm. The generator includes three dorsal swim interneurons (DSI), ventral swim interneurons (VSI) of A and B types, and cerebral interneuron (C2). With the exception of VSI-A, all these interneurons satisfy the fourth step of the strategy. Swimming rhythm could not be changed by stimulation of VSI-A. Getting included it in the generator network nevertheless, but VSI-A does not play a decisive role in the generation of swimming rhythm.

Synaptic connections between these types of neurons are shown in Figure 2. The three DSI are joined by mutually excitatory connections, which are the basis for sustained excitation in the ring (Figure 2a). Connections between other types of interneurons are shown in Figure 2b. A characteristic feature of these monosynaptic responses is their long duration (4–9 sec.).

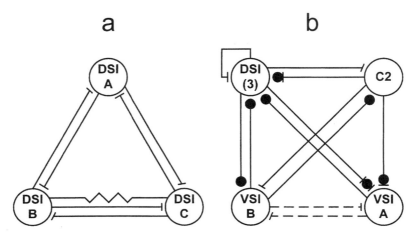

Figure 2—Synaptic connections between neurons of the locomotor generator in the sea slug (according to Getting 1981, 1983b). a, synaptic connections within the group of DSI interneurons; b, connections between different types of generator neurons. Three DSI interneurons are united in one group. Electrical connections are shown as resistors. Monosynaptic chemical excitatory and inhibitory synapses are designated by short lines and dark circles, respectively. Multifunctional monosynaptic connections are shown using combinations of short lines and circles. Dotted lines represent unestablished connections. Explanations are given in the text.

Furthermore, there are multifunctional synapses with, for instance, inhibition followed by excitation between C2 and VSI-A, or excitation followed by inhibition between C2 and DSI (Figures 3e, b). Interaction between the same mediator and different types of postsynaptic receptors is considered to be the basis for multicomponent PSPs (Kehoe 1972a, b). Each receptor type is connected with a definite ion channel, and subsequently de- or hyperpolarization results in different time courses. Four distinct synaptic actions on follower cells could be observed: (1) a fast chloride-mediated IPSP, (2) a fast sodium-mediated EPSP, (3) a slow potassium-mediated IPSP, and (4) a very-slow-conductance-decrease EPSP. Different combinations of these components easily explain the various time courses of PSPs observed.

Investigation of the features of various types of generator interneurons has shown that no one of them was a pacemaker, although VSI-B had a peculiarity. VSI-B responded to depolarizing current by producing a series of action potentials with accelerated frequency. In addition, the first spike was delayed from the beginning of the current pulse. These features are given to VSI-B by a so-called A-current, which is a slowly inactivating K-current (Getting 1983b). A-current is activated during depolarization and explains the behavior of VSI-B.

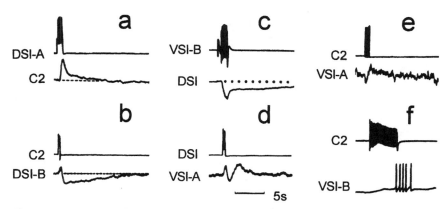

Figure 3—Examples of monosynaptic responses in different interneuronal types found in the sea slug generator (according to Getting 1981, 1983b). In a–e: upper curves, responses of stimulated interneurons; lower curves, responses of neurons monosynaptically connected with the stimulated ones.

The activity of generator interneurons is shown in Figure 4. The cycle begins with the appearance of bursts in DSI that lead to inhibition of VSI-B and to excitation of C2. After a certain delay (necessary for the summation of synaptic influences), C2 begins to fire and to excite VSI-B. A definite time is necessary for depolarization of VSI-B to a threshold level. Excitation evoked by C2 must overcome the inhibition by DSI. VSI-B has quite a high threshold. Moreover, the A-current is activated at the beginning of its depolarization.

Thus, the polysynaptic excitatory chain DSI-C2-VSI-B provides a delayed excitation of VSI-B, and the A-current plays a significant role in this delay. As soon as VSI-B begins to fire, DSI and C2 are inhibited and stop firing. C2 is the main source of excitatory influences on VSI-B, so the cessation of its

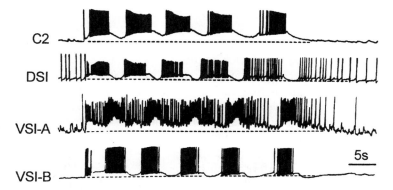

Figure 4—Examples of interneuronal activity during swimming in the sea slug (according to Getting and Dekin 1985).

burst removes this excitation. Trace excitation in VSI-B decreases, and the burst in it disappears. The disappearance of inhibitory influences from VSI-B gives DSI the capacity for new excitation. Therefore, monosynaptic recipro-cal inhibition between DSI and VSI-B, paralleled by polysynaptic excitation of VSI-B from DSI through C2, forms the basis for rhythm generation.

Chain DSI-C2-VSI-A could also be transformed to a chain with recipro-cal inhibition paralleled by polysynaptic asymmetric excitation. In this polysynaptic chain, the necessary delay of excitatory influence on VSI-A is accomplished on the basis of features of synapses from C2 and DSI. There-fore, two networks including VSI-A and VSI-B, respectively, have similar functional meanings. They provide rhythm generation and can enforce each other because of the mutual excitatory influences between VSI-A and VSI-B. However, the network, including VSI-B, obviously plays a more important role in rhythm generation.

It is obvious that constant excitatory influence on such a system is neces-sary for originating a continuous self-generating process. Moreover, the corresponding inputs to the neural network must be organized in a definite way. According to Getting and Dekin (1985), the activating influences come to DSI and VSI-B. In this case, the generator is active during the initiating signal. In its absence, the generator cannot produce the swimming rhythm. Computer simulations based on the above-described neuronal interconnec-tions, cellular and synaptic properties have demonstrated a good coinci-dence between the model and the behavior of the real generator network (Getting 1989).

Clione limacina

The sea angel is usually found in a vertical position. The animal regulates the depth of swimming by making continuous rhythmic movements of its two "wings" in dorsal (D-phase) and ventral (V-phase) directions. The frequency of wing oscillations is usually 1–2 Hz, but it increases 2–3 times during the catching of prey.

The pedal ganglia contain neuronal mechanisms generating the locomo-tor rhythm. Moreover, the swim pattern for each wing can be produced unilaterally. Each pedal ganglion contains about 400 neurons. About 60 of them are rhythmically active during generation of the swimming rhythm. These rhythmically active cells are divided into nine groups: three groups of interneurons and six groups of efferent neurons. The rhythm generator itself includes three groups of interneurons: cells of the 7th, 8th and 12th types (Figure 5) (Arshavsky et al 1985 a–e).

Electric responses in efferent neurons lack any specific features. Depolari-zation of these neurons leads to the appearance of one or several action potentials which can be blocked by tetrodotoxin and are of short duration (about 1–5 ms). Unlike efferent neurons, the generator interneurons have

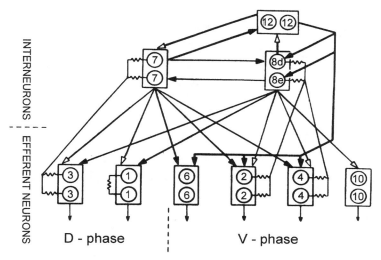

Figure 5—Sea angel's locomotor generator (according to Arshavsky et al 1985d). Chemical excitatory and inhibitory connections are shown by white and black arrows, respectively. Electrical connections are shown as resistors. Neuron types are designated by numerals. Explanations are given in the text.

several peculiarities. Interneurons of types 7 and 8 generate action potentials whose duration ranges between 50–150 ms and cannot be blocked by tetrodotoxin. Interneurons of 12 type are "plateau" interneurons. They generate no action potential. Their membrane potential has two stable states separated by 30–40 mV. Short de- and hyperpolarizing current pulses, as well as excitatory and inhibitory PSPs, can transfer the interneuron from one state to another. Low frequency locomotor rhythm is produced by interneurons of 7 and 8e ("early") types, while interneurons of types 12 and 8d ("delayed") participate in generation of high-frequency rhythm.

Thus, the low-frequency generator is a system including two half-centers that mutually inhibit each other. As was shown above in the case of *Tritonia*, tonic activating signals and delayed excitations are necessary for the stable functioning of this system. Mutual excitatory influences between type 7 and type 8 neurons were experimentally found as well. A rebound mechanism also plays a significant role in rhythm generation. This mechanism is well expressed in these types of neurons. Finally, nonrhythmic neurons whose polarization changed the generator activity were found in pedal ganglia (Arshavsky et al 1984). It was possible to increase or decrease generator rhythm by influencing these command neurons.

The activity of type 7 and 8e neurons and their connections with efferent neurons explain well the low-frequency locomotor pattern (Figure 6a). The type 12 interneuron plays a key role in high-frequency locomotion. Genera-

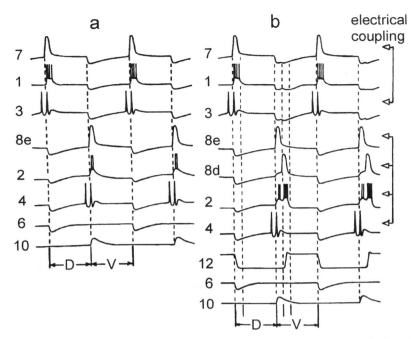

Figure 6—Synaptic and action potentials in different types of neurons during slow (a) and fast (b) swimming in the sea angel (according to Arshavsky et al 1985b, c). For explanations, see text.

tion of action potential in the type 8d interneuron is necessary for activation of type 12 neurons. Type 12 neurons have a double function. One of these functions is negative feedback in relation to V-phase neurons (Figure 5). V-phase neurons are electrically coupled; they mutually excite one another, and a regenerative process is possible in such a system during high levels of activation. The other function of type 12 interneurons consists of acceleration of transition of the type 7 interneuron into the excitatory state after its inhibition by type 8 interneurons. Shortening of the rhythm period is the result of such acceleration.

The pacemaker activity of neurons included in the sea angels system plays an important role in controlling locomotor behavior. This was directly shown in experiments on isolated neurons of different types. Isolated type 7 and 8 neurons spontaneously generated action potentials with frequencies of about 0.5–5.0 Hz. Isolated efferent neurons also produced action potentials, but their frequency was lower. These features are advantageous to the system from the energy point of view. The active generator state is the main characteristic of the sea angel, and it needs minimal influence from command neurons.

After an analysis of generator organization (not locomotor only), Getting (1986) came to the conclusion that a single plan of generator organization is absent in invertebrate species. CPGs do not appear to operate by a single mechanism, but rather by the interaction of many. Network, synaptic, and cellular mechanisms can provide generator activity. CPGs for most continuous behaviors are associated with one or more neurons having pacemaker properties. CPGs that are used rarely and produce only a few cycles do not incorporate pacemaker neurons, but rely on a balance of synaptic excitation and inhibition.

It is obvious that significant increase in the number of elements included in the controlling system limits the applicability of the above-described investigative strategy. For example, in the case of the sea angel, it is possible to see some uncertainties in CPG description; one can find information about connections between different groups of neurons, which explains the rhythm generation, but there is no exact schematic of connections among all the 60 (more exactly, about 60) rhythmically active neurons. It is clear that if someone wanted to obtain the exact schematic of connections among 60 or even 400 neurons in the pedal ganglion of the sea slug, it would indeed be possible from the practical point of view to obtain such information. However, it is doubtful that this additional information would give us a new and better understanding of the mechanisms of rhythm generation in the sea angel.

As we will see below, the limitations of Getting's strategy of CPG investigation become more obvious, when it is applied to vertebrate generators, particularly to generators in higher vertebrates.

Vertebrate generators

In lower and higher vertebrates, the number of elements in the controlling system exceeds, by many orders, the number found in invertebrates. Therefore, the performance of several of the abovementioned steps of investigation becomes not only difficult, but virtually impossible. For instance, at the fourth step, a neuron is identified as belonging to a generator if its stimulation or polarization changes the motor pattern. In vertebrates, each functional cell group is usually quite numerous. Moreover, synaptic weights, as a rule, are so small that during generation of an action potential in one cell, subthreshold PSPs only appear in the target cells. Their amplitude usually does not exceed several tens of microvolts in higher vertebrates. Consequently, accurate identification of the interneuron as belonging to the generator is impossible, because a change in the activity in a single neuron does not cause visible changes in the whole system. It is principally impossible to map all synaptic connections (fifth step of the strategy) in vertebrates. At least, it is impossible to do this using the methods of contemporary neurobiology.

Numerous studies were conducted to investigate the architecture of CPGs in lower and higher vertebrates by using Getting's strategy. Two networks coordinating swimming in lower vertebrates, in the lamprey and in the *Xenopus* embryo (Buchanan 1982, 1986; Buchanan and Cohen 1982; Dale and Roberts 1985, Grillner et al 1986; Roberts and Tunstall 1990; Roberts et al 1981, 1986; Rovainen 1974, 1979, 1986; Soffe 1990; Soffe et al 1984; Soffe and Roberts 1982a, b, 1989), have been described in considerable detail. Less-detailed information was obtained about the architecture of CPGs for locomotion and scratching in higher vertebrates: in cats (Baev et al 1979, 1980, 1981; Berkinblit et al 1978; Edgerton et al 1975, 1976; Jankowska et al 1967; Orlovsky and Feldman 1972) and in rabbits (Viala and Viala 1977).

Lower vertebrates

Specific muscle activity is inherent in swimming. Myotomes are controlled by corresponding spinal cord segments. Contraction of myotome muscles on one side is accompanied by their relaxation on the other side. The spinal cord provides (with a time lag) the rostrocaudal activation of body muscles during swimming.

Several different classes of spinal interneurons participating in the generation of the locomotor rhythm in the lamprey have been identified, along with some types of sensory control of the network, and the descending pathways for initiation of locomotion. The nature of synaptic connectivity and the types of transmitters involved in it have been established by experiments with paired intracellular recordings. Figure 7 shows a conceptual model of the segmental rhythm-generating network that was proposed on the basis of these experimental findings (Grillner et al 1996). The network interneurons and motoneurons are subject to a descending excitatory drive from brain stem reticulospinal (R) neurons. The network neurons also receive inputs from intraspinal stretch-receptor neurons (SR) called edge cells, activated by locomotor movements.

The capacity of the network to generate an alternating rhythm can be understood in terms of the established connectivity between the three types of interneurons on each side of a segment: the excitatory interneurons (E), the lateral interneurons (L), and the crossed inhibitory interneurons (I). The motoneurons (M) are considered by the authors as output elements that are not a part of the rhythm-generating circuitry. During natural conditions, the generator is activated by the bilateral excitatory drive from the brain stem, which produces a high level of background excitability. The active side keeps the contralateral side silent because of reciprocal inhibition via I neurons. It is seen from the network scheme that the activity of L neurons may serve as one of several burst-terminating mechanisms. The L neuron fires late during the ipsilateral burst and inhibits, the I neuron, and consequently the network neurons on the contralateral side become disinhibited. This will result in activation of the contralateral side and inhibition of the previously active side. Obviously, rather complex synaptic and cellular mechanisms also play

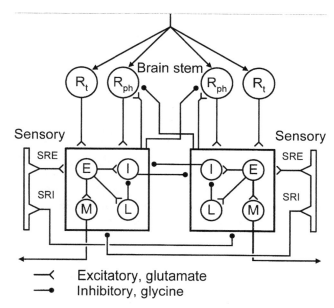

Figure 7—A conceptual model of a segmental network generating locomotor rhythm in the lamprey (according to Grillner et al 1995). Neuron symbols denote populations of neurons rather than single cells. All the spinal network neurons depicted within the box are excited by glutamatergic reticulospinal tonic (R_t) and phasic (R_{ph}) brain stem neurons. The excitatory interneurons (E) excite all types of spinal neurons within the box. The inhibitory glycinergic interneurons (I), whose axons cross the midline, inhibit all the neurons of the contralateral box. Lateral interneuron (L) inhibits I interneuron. Motoneurons (M), which are cholinergic, send signals to the muscles. Stretch-receptor neurons are of an excitatory (SRE) type that excites neurons within the ipsilateral box, and of an inhibitory (SRI) type that inhibits the neurons within the contralateral box.

an important role in rhythm generation. One of them is, for example, the pacemaker property of lamprey spinal neurons.

During swimming, incoming sensory information from the stretch receptors, edge cells, can adjust the generator activity. When one side of the body is contracted, the other one is extended, and the latter triggers the stretch receptors. These activated nerve cells have a double effect on the generator. First, they excite neurons on the extended side, which results in contraction of corresponding muscles. Second, they inhibit neurons on the contralateral side, which stops contraction on that side.

Intersegmental interaction during swimming is based on the specifically organized excitatory and inhibitory influences of one segment on other segments of the spinal cord. Local neural networks extend axons along the spine. Special inhibitory cells of one segment send signals through these axons in the caudal direction for as much as one-fifth of the length of the spine. Excitatory cells have shorter axons that project in both directions.

Numerous computer simulations mimicking the lamprey spinal neural network were conducted (see Grillner 1996; Grillner et al 1996). Those simulations are probably the most extensive ones among all the simulations that have been made by now in studying the CPG problem. The authors included in their model detailed data about the spinal network architecture, and synaptic and cellular properties. These simulations demonstrated that the computer model can account for essential features of the motor pattern seen experimentally during lamprey locomotion. Moreover, the authors even succeeded in modeling the entire lamprey, "from the muscle fibers controlling the different segments to the viscous properties of the surrounding water". Virtual swimming by a simulated lamprey can portray how the real creature moves through the water.

CPGs in lamprey and *Xenopus* embryos (Roberts et al 1995) appeared to be very similar (Figure 8). However, the latter has its own specific burst-initiating and burst-terminating mechanisms that are probably connected with a much higher frequency of swimming movements in this animal.

These studies of CPG organization in lamprey and *Xenopus* embryos are the most detailed among those that were conducted in vertebrates and

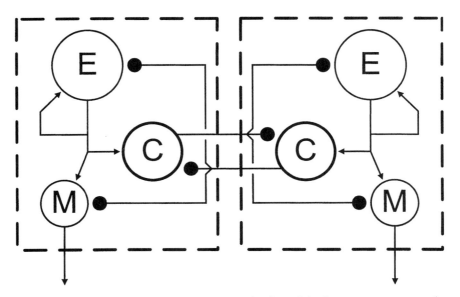

Figure 8—A conceptual scheme of the organization of the locomotor generator in the *Xenopus* embryo (according to Roberts et al 1986). E and C, excitatory and contralateral inhibitory interneurons, respectively; M, motoneurons.

based on the mechanistic approach. However, as can be well seen, the description of these CPGs was not presented in the form of a complete scheme of connections between neurons; the reconstruction of the generator network was based on the conceptual model of major connections explaining rhythm generation. It is still unknown how exteroceptors and other descending commands influence the generator activity.

Higher vertebrates

In higher vertebrates, such as cats, increments of ten muscles are involved in the hind-limb locomotor and scratching rhythmic movements. It is common to subdivide the locomotor cycle into flexion and extension phases. The scratch cycle is subdivided into aiming and scratching jerk phases. There is also an initial aiming period during scratching, when the tip of the hind limb moves towards the pinna. Numerous studies have shown that the locomotor and scratching programs for the hind limb are very complex, both during real and fictive locomotion and scratching. Commands to some muscles (for example, double joint muscles and some distal muscles) arrive in both phases, and there is a substantial coactivation of antagonistic distal muscles (Figure 9).

Obviously, recording from single spinal moto- and interneurons revealed changes in their activity during fictive movements. Alpha- and gamma-motoneurons are rhythmically active during fictive locomotion and scratching. Some rhythmic interneurons are active only in one phase, while others fire in both phases. Interneurons that tonically change their activity during fictive locomotion or scratching have also been found (Figures 10, 11). Not all of the investigated neurons have been properly identified. The majority of interneurons that change their activity during CPG activity receive inputs from various peripheral afferents, but strict correlation between the type of afferent input and the pattern of interneuronal firing has been revealed only for some types of interneurons. For instance, Ia interneurons fire in phase with their motoneurons. The same is true for some premotor interneurons. One of the most important experimental findings is the fact that the same spinal interneurons are active both during fictive scratching and locomotion, which means that the same interneurons are included in both CPGs. Finally, some rhythmically active interneurons were identified as interneurons of various ascending tracts. These results will be discussed in Chapter 10.

Hypotheses of rhythm generation

Current views of the spinal locomotor or scratching CPGs in higher vertebrates exist only in the form of hypotheses. Numerous such hypotheses have been proposed, and presently they have only a historical interest. It is necessary to point out that according to M. L. Shik (1976), a well-known

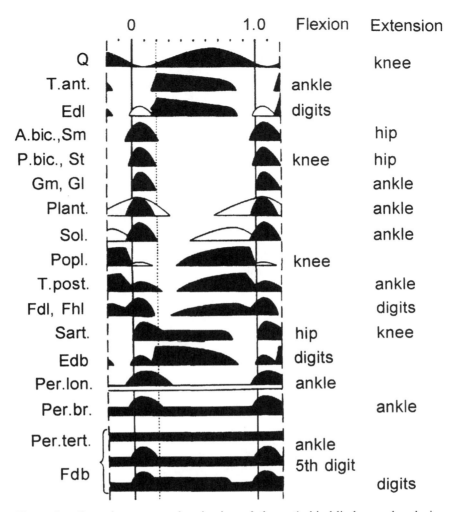

Figure 9—Central program of activation of the cat's hind-limb muscles during fictive scratching. Black and white figures designate frequent and rare forms of activity in different nerves, respectively. The muscle abbreviations are shown on the left; on the right, functions of corresponding muscles. Abbreviations: Q, m. quadriceps; T.ant., m. tibialis anterior; Edl, m. extensor digitorum longus; A.bic., m. anterior biceps; Sm, m. semimembranosus; P.bic., m. posterior biceps; St, m. semitendinosus; Gm, m. gastrocnemius medialis; Gl, m. gastrocnemius lateralis; Plant, m. plantaris; Sol., m. soleus; Popl., m. popliteus; T.post., m. tibialis posterior; Fdl, m. flexor digitorum longus; Fhl, m. flexor hallucis longus; Sart., m. sartorius; Edb, m. extensor digitorum brevis; Per.lon., m. peroneus longus; Per.br., m. peroneus brevis; Per.tert., m. peroneus tertius; Fdb, m. flexor digitorum brevis. Notice that activity in nerves to some muscles, such as Per.tert. and Fdb, varied in different experiments.

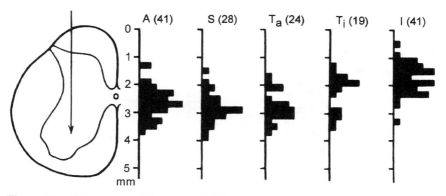

Figure 10—Histograms of location of different types of interneurons on the transverse cut of the spinal cord. A, aiming interneurons; S, scratching interneurons; T_a, tonically active interneurons; T_i, tonically inhibited interneurons; I, indifferent interneurons. Number in parentheses is the number of investigated neurons.

specialist in neural control of locomotion, "all these hypotheses have one common demerit—inconcreteness." However, we will briefly discuss some of them below because they demonstrate extremely well the level of theoretization in the CPG problem.

Brown's hypothesis, which was later developed by Jankowska et al. (1967), is the oldest one (Figure 12a). Brown (1914) proposed this hypothesis based on the observation that alternating flexion and extension movements can occur in a deafferented animal after spinalization. According to this hypothesis, the locomotor program is the result of the alternating activity of two half-centers, the flexor and extensor half-centers. According to Brown, the reverberation process started in one half-center terminates because of accumulation of fatigue in it. Another half-center becomes active as a result of this fatigue. Activity-terminating mechanisms are the least understandable and the most essential in Brown's hypothesis. In addition, even superficial analysis shows that it is hard to explain the complex locomotor program by means of this hypothesis, unless someone postulates that both half-centers activate motoneurons of those muscles that are active in both phases.

Miller and Scott's (1977) hypothesis accounts for concrete data about connections of Ia interneurons, motoneurons, and Renshaw cells (Figure 12b). It proposed that half-centers can be built of Ia interneurons. The oscillatory process in such a system occurs as the result of recurrent inhibition of the active half-center. However, this model cannot explain such experimental facts as the appearance of rhythmic processes in the absence of motoneuronal activity, and this circuitry should be considered as a part of the generator.

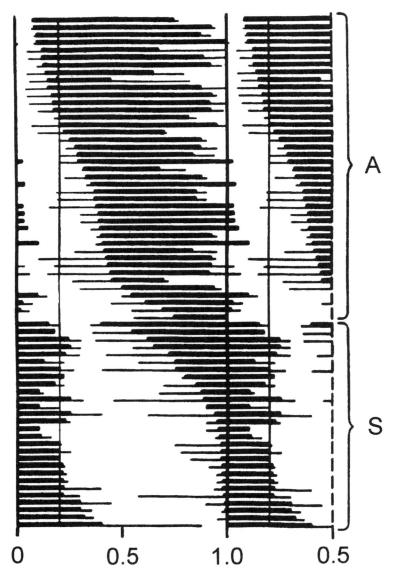

Figure 11—Distribution of phases of activity of aiming and scratching interneurons in the normalized scratch cycle. Thin lines, periods of increased neuronal activity; thick lines, periods during which an increase of neuronal activity was higher than 0.5 of the frequency modulation; A and S, aiming and scratching interneurons, respectively.

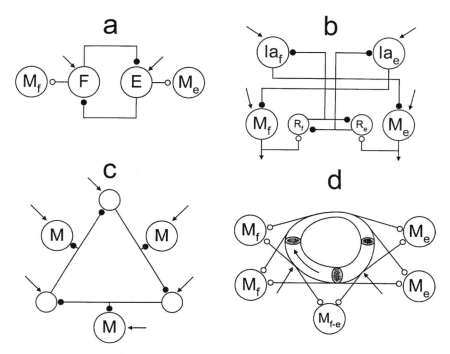

Figure 12—Hypothetical models of central pattern generators. a, Brown's model; F and E, flexor and extensor half-centers; M_f and M_e, flexor and extensor motoneurons; b, hypothesis of Miller and Scott. Ia_f, Ia_e and R_f, R_e, Ia interneurons and Renshaw cells of flexors and extensors; c, d, models of single and multilayer rings, respectively; M_{f-e}, motoneurons of double-joint muscles. Arrows designate tonic excitatory influence.

Half-center hypotheses are a particular case of later hypotheses that described a generator as a ring of interneurons in which excitation circulates. This ring (Figure 12c, d) can be a linear chain of neurons (Szekely 1968; Kling 1971) or can be multilayered (Gurfinkel and Shik 1973; Shik 1976). Motoneurons are connected to different parts of the ring that determine their sequential activation in the motor cycle. Obviously, different cellular and network mechanisms can regulate the direction and the spreading of excitation along the ring. These mechanisms can effect temporal and spatial summation. The number of layers can also play a significant role in the spreading of excitation along the ring. The excitation spreads faster through a thick part of the ring. The number of layers can be regulated by descending commands. Clearly, views of a generator as a multilayer ring can reasonably explain swimming in fishes (see above). In terrestrial mammals, a spreading of an excitation wave along the spinal cord was not described. Therefore, one could conclude that the circulation of excitation can occur along the trajectory that is not oriented along the spinal cord.

There are also views that a generator can be built on the basis of several groups of interneurons that are specifically connected. For example, the spinal scratching generator was hypothesized as having this construction (Berkinblit et al 1978). Some hypotheses explained generator rhythmic activity as being determined by the activity of pacemaker neurons, but these hypotheses were mainly related to generators in invertebrates and, maybe, lower vertebrates.

None of the generator hypotheses excludes the role of peripheral feedback. However, its role is secondary in relation to rhythm generation. Peripheral feedback is necessary for correction of generator activity.

Another group of hypotheses puts the role of feedback in first place. Walking has been considered a chain reflex, and foot receptors and proprioceptors play the major role in it (Philippson 1905; Sherrington 1910a). Magnus (1924, see Magnus 1989) proposed to consider walking as the means to support equilibrium; if a limb is in a position in which it cannot support the body, it performs a stepping movement. Therefore, it is possible to designate this group of views as afferent generator hypotheses. Clearly, these hypotheses cannot explain rhythm generation without feedback.

Thus, it is clear that all the research in generator problems, both experimental and theoretical, has been dedicated to understanding the mechanisms of rhythm generation. Some researchers succeeded in doing this for several particular simple cases. However, one has to admit that all the experimental and theoretical results have shown that there is no single plan of generator construction, neither in invertebrates nor in vertebrates. In fact, one simple truth was discovered: if different body parts are working in synchrony during automatic motor behavior, or partly in synchrony, there is usually mutually excitatory influence between the neural circuits governing those movements. If the body parts are working out of phase, relations between corresponding networks are antagonistic and they inhibit each other. This is exactly what was first observed and proposed by Brown. It is necessary to mention that even those generators that are considered to be understood are still within the framework of Brown's hypothesis; two mutually inhibiting half-centers can be found in them.

The results of the research discussed above led many scientists to conclude that the state of the generator problem was critical. A crisis came up in connection with the enormous complexity of the generator networks in higher vertebrates. It became obvious that the methods used for investigation of motor control systems in invertebrates were inadequate for the study of analogous tasks in higher vertebrates. Moreover, even notions of "incognition" in generators of automatic movements appeared (Selverston 1980).

1.1.4 Afferent Correction of Central Pattern Generators

As was already mentioned in Section 1.1, the combination of the program control principle with the reflex theory has been the theoretical basis for investigations in the field of generator afferent control for the last several decades. The appearance of such a theoretical basis is a natural consequence of the previous developments in neurobiology. Let us examine this problem, using the example of higher vertebrates.

Generator interneurons and motoneurons receive inputs from the peripheral receptors—for example, from the limb—directly or through relay neurons. Moreover, inputs are organized in such a way that the same type of afferent signals can arrive in the neurons working in different phases (Figure 13). Thus, neurons changing their activity during generator work are also

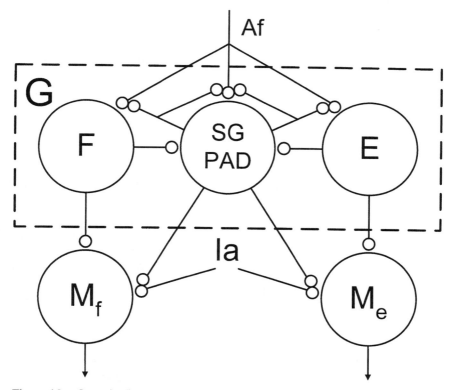

Figure 13—Organization of afferent flow to generator half-centers in mammals. G, generator; Af, afferent inputs; SG PAD, system generating primary afferent depolarization; F and E, flexor and extensor half-centers (reciprocal inhibitory connections between them are not shown). M_f and M_e, flexor and extensor motoneurons; Ia, Ia afferents.

included in the reflex arcs. In addition, the generator influences the PAD generating system (Baev 1979, 1980; Baev and Kostyuk 1981, 1982). Therefore, any change in the arcs will inevitably lead to a change in reflex responses.

It is clear from the scheme just presented that complex interaction between central and peripheral processes takes place during rhythmic movement control. On one hand, efferent signals change the state of the limb, and, as a consequence, there is a change in the afferent flow from the peripheral receptors. On the other hand, this flow influences the state of the spinal neurons. The existing theoretical basis, and the ensuing mechanistic experimental strategy, directed experimenters towards multilateral investigation of this complex interaction. The following points were investigated:

1. Tonic changes in segmental apparatus and reflexes during activation of generators
2. Phasic changes of segmental apparatus and reflexes during rhythmic generator activity
3. Activity of various types of peripheral receptors during rhythmic movements
4. Changes of generator activity under the influence of peripheral signals

It was found that tonic and phasic changes are observed in all parts of reflex arcs, from input to output (Baev 1981a, b, 1991; Grillner 1975; Grillner and Wallen 1985; Shik 1976; Shik and Orlovsky 1976). The central influence on reflex responses was submitted to a quite simple rule: transmission of peripheral signals through active neurons was facilitated, and it was depressed through inhibited ones (see, for instance, Figure 14).

Research into center influences on the reflex responses has usually been conducted in experimental conditions such that afferent testing stimulus does not evoke significant changes in the generator's work. The experimenter uses more powerful afferent stimuli for investigation of regularities of changes in generator work under the influence of peripheral signals. These "disturbing" influences, being applied in different phases, evoke different effects. In comparison with the effects of testing stimulation, these effects are significantly less studied.

A "disturbing" influence can lengthen or shorten one of the phases of generator activity, or even break it, and it can also alter the intensity of efferent activity. Regularities observed could be partially explained by afferent input organization. For example, signals coming from flexor reflex afferents at the beginning of flexor phase lengthen that phase, while those relayed via low threshold cutaneous fibers influence mainly extensor activity. Flexor reflex afferents arrive mainly in flexor half-center, and low threshold cutaneous fibers arrive mainly in extensor half-center (Baev et al 1979, 1980).

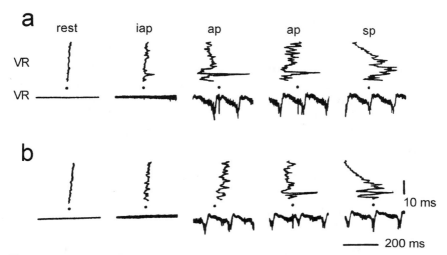

Figure 14—Dependence of monosynaptic response on the phase of fictitious scratching. a and b, stimulation of nerves subserving m. tibialis anterior and m. gastrocnemius, respectively, with the strength of 1.7 threshold. VR, recording from L_7 ventral root; iap, initial aiming period; ap and sp, aiming and scratching jerk phases, respectively.

Finally, entraining phenomena must be picked up from other effects of the "disturbing" stimulation. Synchronization between generator work and stimulating frequency, near to the frequency of generator activity, is usually observed. The physiological role of this regularity is obvious, but the neuronal mechanisms underlying it are not clear.

Numerous works have demonstrated that the receptors of rhythmically moving parts of the body (limb) are rhythmically active. The character of the receptor activity depends on its type, localization, and other factors.

Therefore, the next phenomena are now established: tonic and phasic reflex dependence, rhythmic activity of receptors, and entraining phenomena.

It is not expedient to perform more detailed analyses of correction mechanisms based on the mechanistic approach. Almost all data obtained up until now have a mainly qualitative character, and their detailed analysis, not including the abovementioned, would require the publication of a very voluminous book. Moreover, as follows from the previous description, the current strategy of investigation of the correction problem requires that the problem of CPG and reflex arc organization be solved first which has not been done to date. In this connection, it could be concluded that the current strategy of brain investigation, based on reflex theory and the program control principle, has limited application to the solution of the problem of peripheral correction of generator motor activity.

1.1.5 Ontogenesis of Locomotor Function

Locomotion is one of the functions that is well developed by the moment of birth in the majority of animals. How does it occur? Is the wiring of the generator network completely predetermined by the genetic mechanisms, or do some forms of learning have to take part in the process of network development? In discussions of the CPG problem, these questions are rarely raised, and the majority of scientists think that generator networks are genetically prewired. However, some data are presently available in the scientific literature that indirectly show the importance of learning during generator development. This information and a discussion of its significance will be presented in Section 8.5, because current mechanistic views on the generator problem are not capable of explaining it.

1.1.6 Is the Formulation of the Generator Problem Correct?

The higher we move along the evolutionary scale of vertebrates, the more clear it becomes that, from the methodological poin of view, the popular mechanistic approach to biological neural networks is going to fail. It is not feasible yet to examine the neural circuits in mammals in much detail by using this approach.

Within the limits of the current neurobiological strategy of motor automatism investigation, the motor pattern generator is considered from a purely metaphysical position, as something from Heaven, preprogrammed from the moment of the animals birth. Peripheral correction is considered from the same position, i.e., as preprogrammed dependence of numerous reflexes on the generator state. All this implies the recoding of input signals to output ones, which is postulated by reflex theory, but with one difference: the addition of dependence on the generator activity phase. All the drawbacks of reflex theory are preserved using such an approach. Reflex theory initiated the investigation of all the types of reflexes. In addition to this variety, it is also necessary to investigate reflex dynamics and organization of CPGs, applying reflex theory to the investigation of inborn automatic behaviors. It is clear from the above discourse that reflexes and generators vary in different animal species, and that the ability to elucidate principles of brain functioning by means of detailed investigation of countless reflexes and generators is doubtful. To date, the modern trend in neurobiology has been to go in just this direction. At best, what could be obtained using this approach is information about the work of a simple neural network, but not about why it is organized exactly as it is. The same limitations are applicable to any synthesis based on a mechanistic approach. Presently, this method relies on the detailed computer simulation of the circuitry being studied. It does not matter how much time is spent on a sophisticated simulation of the concrete network activity or even the whole motor behavior. It will always be a simulation that is not capable of revealing the principles of system construction, because such a simulation is designed to mimic the external behavior of the investigated system and can be designated as phenomenological. This is the major reason that such synthesis has a limited generalizing power, and nobody has yet succeeded in reaching more general conclusions by doing detailed simulations of simple systems.

Using motor automatic behaviors as an example, we have confined ourselves to the analysis of generators for rhythmic movements and the mechanisms of their peripheral correction. We have not considered the role of higher brain areas in the control of those behaviors, in particular the role of suprasegmental correction. This question will be discussed in Chapters 10 and 11. We note here only that the current strategy of investigation of these questions is similar to the strategy used in peripheral correction investigation.

Current strategy in generator investigations has one additional aspect. Up until now, as has been demonstrated, efforts have been directed mainly at investigating rhythm mechanisms. Generator influences on efferent neurons are usually simple in simple systems, i.e., a coordination problem solved by the system is quite simple. That is why discoveries about the mechanisms of locomotor control could be applied to ascertaining the mechanisms of rhythm generation. The coordination task is at the forefront in complex systems. For instance, in mammals, as mentioned above, increments of ten muscles evoke limb movements, and their program of activation during scratching and locomotion is complex. Hence, current strategy is not applicable to the elucidation of the coordination task. As we also saw above, it appeared to be ineffective even for discovering the rhythm mechanisms in higher vertebrates.

In concluding the analysis of the current state of the CPG problem, it is necessary to focus the reader's attention on experimental facts that are not satisfactorily explained within the limits of the contemporary theoretical model (they are usually not discussed during consideration of the problem).

1. The CPG is able to generate the motor program without peripheral feedback, or after a partial break in afferent information channels. Moreover, the CPG always generates a rhythm that is very close to the rhythm observed in an intact animal. It is obvious that the necessary rhythm depends on the physical characteristics of the moving body part. If these characteristics change (for example, during ontogenesis), then the rhythm has to change. How does the generator adjust its rhythm during ontogenesis? It is highly improbable that this process is predetermined genetically.
2. The distribution of the generator system and its possession of holographic features, mean that the system can function even after its par-

tial destruction (as in holography, in which a complete three-dimensional image of an object can still be reproduced after partial destruction of the photographic plate, but with a lesser resolution). This is most prominent in the complex systems. In vertebrates, even a small part of the generator neural network (for instance, one or several spinal cord segments) can produce locomotor rhythm. In invertebrates, each type of neuron included in the generator is usually represented by several cells, and the loss of one of them does not abolish the generator activity.

3. Neural structures serving as the basis for generator formation under the influence of the initiating signal are usually involved in the production of other motor programs. The spinal neural network of mammals, used for postural control, scratching, locomotion, and many other movements, is a typical example. Locomotor and scratching generators have many common neurons. Thus, each generator is a different regime of activity of this neural network. The question arises: How is it possible to interpret the information about neuronal interconnections and synaptic and neuronal properties in the case of a multifunctional network?

Therefore, it is necessary to reconsider the meaning of the term "central pattern generator." Is this meaning anatomical or functional? It is clear that the current mechanistic strategy of CPG investigation implies that the term "generator" designates a definite structure, and that scientists have to study it. The above analysis shows convincingly that the term CPG has a functional meaning. Otherwise, one would have to suggest that different structures exist within a structure, such as scratching and locomotor generators in the mammalian spinal cord. It is impossible to consider scratching and locomotion as different regimes of work of the same generator. It has been shown experimentally that scratching and locomotion are antagonistic types of movements, they exclude each other. Obviously, if the analyzed networks are not unifunctional, then it is highly doubtful that the mechanistic approach is applicable to such cases, because it is not clear how to make a correlation between the neuronal interconnections and their function—i.e., how to approach the relationship between structure and function.

Finally, one has to make quite an unpleasant conclusion: The term CPG, accepted from the technical field, where it usually possesses a structural meaning, has been driving the above-described mechanistic strategy of CPG investigation for many years—and this research, based on a wrong assumption, has led us nowhere. Accumulation of a substantial amount of experimental data was the only positive result of all previous research. However, this accumulated information is waiting for the correct explanation. We shall see how the same data can be explained using a different approach, in which CPG has only a functional meaning.

1.2 Learning from the Point of View of Classical Theory

As a rule, one of two conditioned reflex models is used in the fields of experimental neurobiology where mechanisms of memory and learning are investigated. It is customary to distinguish two types of conditioning—classical and instrumental conditioning. Classical conditioning is frequently characterized as the associative conditioning of Pavlov. Here, a conditioned reflex appears after the presentation of conditioned (CS) and unconditioned (US) stimuli in close temporal association with each other. Temporal and spatial relations between these stimuli are significant factors to be accounted for in learning. A commonly used temporal parametric definition of classical conditioning is shown in Figure 15. It is noteworthy that the most efficacious conditioning paradigm is simultaneous conditioning, in which the CS terminates at the end rather than at the beginning of the US (Poulos et al 1971).

In the instrumental type of conditioning, the probability of the response is altered by the contingent occurrence of reinforcing or punishing stimuli.

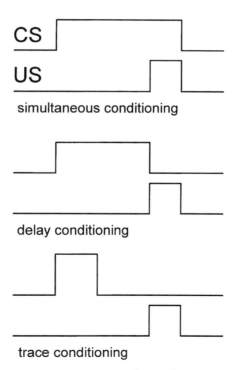

Figure 15—Parametric definition of classical conditioning. CS and US, conditioned and unconditioned stimuli, respectively.

It is the response of the animal's environment upon delivery, production, or cessation of the stimulus that is crucial. This provides the basis for an associatively induced adaptation. Therefore, ordinarily, during instrumental conditioning, the response of the animal influences the probability of occurrence of reinforcing or punishing stimuli.

To many, the physiological basis of instrumental conditioning appears to be equivalent to escape or avoidance conditioning, but that is not completely true. Typical contingencies for instrumental reflexes are shown in Figure 16. In the present section, we restrict ourselves to the consideration of several experimental facts obtained through the models of classical conditioning. In Chapter 12, the principal differences between these two types of conditioned reflexes are shown.

At present, the phenomenology of conditioned reflexes is extensively detailed, and intensive work continues in this field. But, as was discussed earlier,

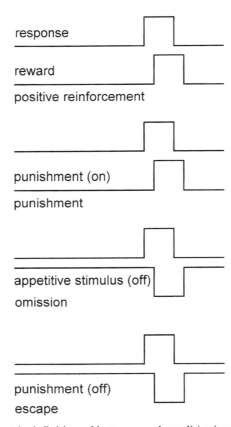

Figure 16—Parametric definition of instrumental conditioning.

the analytical mechanistic strategy of brain investigation dominates modern neurobiology, and therefore the emphasis in the investigation of mechanisms of conditioned reflexes is on the search for corresponding cellular mechanisms. Moreover, the same tendencies observed in the generator problem, are evident: experiments are conducted on simple subjects, invertebrates, cells *in vitro*, etc. Some cellular and synaptic changes during conditioning have been carefully documented in higher invertebrates (Carew and Sahley 1986; Byrne 1987; Edelman et al 1987; Kandel et al 1992; Siegelbaum et al 1991).

The main idea of experiments at the cellular level is rather simple. These experiments are based on the hypothesis that memory is associated with use-dependent synaptic modifications (see Brown et al 1990). When this hypothesis is applied to conditioned reflexes, it has been suggested that a new reflex arc (or a modification of an old one) appears as the result of the establishment of a new association between stimuli—conditioned and unconditioned. This is why it is believed that changes in the efficacy of synaptic transmission should occur and scientists try to find the locations of these changes, i.e., they try to find memory traces.

Hebb's rule (1949) is the typical example of current views on use-dependent synaptic modification: "When an axon of cell A is near enough to excite cell B or repeatedly or consistently takes part in firing it, some growth process or metabolic change takes place in one or both cells such that A's efficiency, as one of the cells firing B, is increased." It has been suggested that this rule operates in higher vertebrates.

Even superficial analysis of this rule shows that it does not possess the necessary functional completeness to explain the existing experimental facts. For instance, Hebb's rule cannot explain the experimentally established fact that there is a decrease in synaptic transmission after learning. The direction of changes in synaptic transmission can be determined by the experimenter. Operant-conditioned plasticity in the spinal cord (Wolpaw and Herchenroder 1990; Wolpaw et al 1991) is a typical example. The spinal stretch reflex has been conditioned in the biceps brachii of monkeys. Two types of conditioning were used: "up-mode" and "down-mode." In the first case, reward followed only if the H-reflex in one leg (the conditioned leg) was above a criterion value and in the second case below it.

Hebb did not consider the case of uncorrelated or negatively correlated pre- and postsynaptic activity. This was done later by his followers who generalized Hebb's original concept to include a combination of synaptic enhancement and some type of activity-dependent synaptic depression. Such a generalized Hebbian synapse increases its strength with correlated pre- and postsynaptic activity, and decreases its strength with negatively correlated activity.

Many scientists have believed that the Hebbian rule (more exactly, the Hebbian rule in its generalized form) possesses the necessary completeness to explain various forms of learning in the brain. Numerous theoretical studies have been conducted in which authors tried to show that useful and potentially powerful forms of learning and self-organization can occur in networks of elements in which synaptic connections may undergo various Hebbian modifications. These studies have been analyzed in the review of Brown et al (1990).

As we will see in Chapter 6, the idea of reducing learning to a set of cellular rules (especially deterministic rules) is wrong. The discussion of this topic cannot be conducted in this section, because it requires an introduction of some new concepts (see Chapters 2 and 6). However, this position does not contradict the fact that some self-organization can occur in accordance with simple local rules.

1.2.1 Classical Conditioning of the Eyelid Closure Response

At present this experimental model is very popular, and is used in various laboratories. It is considered to be a "model paradigm of associate learning and memory" (Thompson 1989). Typical time relations between the conditioned (tone) and unconditioned (air puff to the eye) stimuli and the conditioned response are shown in Figure 17. Any other type of stimulus, for instance, an air puff to the back, can be used as the conditioned stimulus. Before learning, the eyelid closure response may only be evoked by an unconditioned stimulus. The learned response develops in such a way that the eyelid closure is maximal at the time of the appearance of the unconditioned stimulus.

The search for the memory trace circuit for this form of learning is performed in full accordance with the generally accepted mechanistic neurobiological strategy described above. Two polar points of view on the neurological location responsible for the eyelid closure reflex have resulted from such investigations. The first: memory trace circuits are located in the cerebellum. The second: the cerebellum is not necessary and sufficient for the classical conditioning of this reflex system. Let us consider the corresponding experimental facts.

Main experimental facts supporting the first viewpoint.

1. Decorticate and decerebrate (Norman et al 1977; Oakley and Russel 1972) mammals can learn the conditioned eyelid response. Animals that were first trained and then acutely decerebrated robustly retained the learned response (Mauk and Thompson 1987). This is why these authors concluded that the memory trace circuit is located below the level of the thalamus.
2. Neuronal activity exhibiting the requisite memory-trace properties has been discovered in some regions of the cerebellar cortex and the inter-

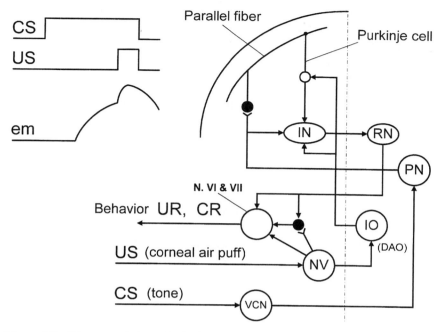

Figure 17—Thompson's hypothetical schema of the memory trace circuit for eyeblink conditioned response (according to Thompson 1989). Conditioned and unconditioned stimuli and eyelid conditioned movement (em) are shown on the left. IN, interpositus nucleus; RN, red nucleus; PN, Pontine nucleus; IO, inferior olive; DAO, dorsal accessory portion of the inferior olive; NV, spinal fifth cranial nucleus; VCN, ventral cochlear nucleus; N. VI & VII, sixth and seventh cranial nuclei.

positus nucleus, in the course of mapping the brain stem and the cerebellum. These neurons showed patterned changes in discharge frequency that correlated with the behavioral learned response and preceded it by several tens of milliseconds (McCormik et al 1981, 1982; Thompson 1989).

3. Presentations of the unconditioned stimulus alone evoke a phasic increase in the responses of dorsal accessory olivary neurons. When the conditioned response begins to develop, the responses of olivary neurons become markedly attenuated. In trained animals, during trials in which the animal gives a conditioned response, activity evoked by the unconditioned stimulus may be completely absent (Sears and Steinmetz 1991). Corresponding changes in complex spike activity in the appropriate Purkinje cells have also been observed (Foy and Thompson 1986; Swain and Thompson 1993).

4. Lesion of the ipsilateral cerebellar circuitry (from mossy fiber input to output—e.g., the interpositus nucleus) abolished the ipsilateral learned response completely and permanently, and had no effect on the unconditioned reflex response. These lesions did not prevent learning on the contralateral side (Clark et al 1984; Lavond et al 1981, 1985; Lincoln et al 1982; McCormick et al 1981, 1982; Yeo et al 1985). According to other authors, however, the unconditioned response may change after lesion of the cerebellar network (Bracha et al 1994).
5. Electrical microstimulation of the mossy fiber system serves as a very effective conditioned stimulus. Moreover, it produces rapid learning—on average more rapid than with peripheral conditioning stimuli (Steinmetz et al 1985).
6. Stimulation of the dorsal accessory olive (climbing fiber stimulation) can elicit a wide range of behavioral responses: eyelid closure, limb movement (flexion or extension), turns of the head, etc. The location of the stimulating microelectrode determines the type of behavioral response. Forward pairing of mossy fiber stimulation as a conditioned stimulus and climbing fiber stimulation as an unconditioned stimulus yields normal behavioral learning of the response elicited by climbing fiber stimulation (Steinmetz et al 1985). Lesion of the interpositus nucleus abolishes both conditioned and unconditioned responses in this paradigm.
7. The effect of electrolytic lesion in the rostromedial (face) region of the dorsal accessory olive differed in untrained and trained animals (McCormick et al 1985). If the lesion was made before training, animals were unable to learn the conditioned response. If the lesion was made after training, the animals showed normal behavioral conditioned responses at the beginning. Later, with continued paired training, the conditioned response was extinguished in a manner very similar to that seen in tests in which the unconditioned stimulus was discontinued and animals were given the conditioning stimulus alone.

The schema shown in Figure 17 is the result of such investigations (Thompson 1989). Nerve impulses evoked by the unconditioned stimulus are conveyed to the cerebellum through the dorsal accessory portion of the inferior olive and its climbing fiber projections to the cerebellum. The conditioning stimulus (tone) pathway consists of auditory projections to the cerebellum. The conditioned response pathway includes the interpositus and the red nuclei. The descending rubral pathway acts directly on the motor neurons. It has been suggested that memory traces are stored in the cerebellar cortex and possibly in the interpositus nucleus. The data obtained from the reduced preparation (see point 6) are the best proof of this. Within the limits of this scheme, the signal from the inferior olive is treated as a teaching signal. However, there have been no suggestions about the possible mechanisms by which learning takes place in the cerebellum or the interpositus nucleus.

An experimental fact supporting the second viewpoint.

Experiments on decerebrate-decerebellate animals (Kelly et al 1990) have demonstrated that these animals were able to learn the conditioned eyelid response independently of whether conditioning had occurred before or after the cerebellectomy. These findings led the authors to conclude that the "cerebellum should no longer be considered as the structure which is necessary and sufficient for the classical conditioning of this reflex system."

Therefore, in the case of associative learning we are facing the same problems that we faced when we analyzed the CPG problem. The mechanistic approach does not answer the following questions:

How are different reflex arcs created on the same anatomical substrate? As we saw, different types of signals can be used as the conditioned stimulus.

How should we treat parallel circuitries?

And the most important question: What are the mechanisms that are responsible for the specific timing of a conditioned response? This question cannot be answered by relying on the simple assumption that an unconditioned response becomes associated with a conditioned stimulus.

2

The Control Theory Approach to Biological Neural Networks

2.1 A Brief Historical Review of the Development of Automatic Control Theory

An exact definition of the concept of "automation" in the broad sense does not exist. This term is usually defined as an apparatus with the ability to work without human intervention. The creation of calculating and regulating automatons has quite a long history (see Baev and Shimansky 1992). The idea of digital apparatus construction was elaborated as far back as the eighteenth century by Pascal and Leibniz. The first works on control system theory appeared in the nineteenth century. In Maxwell's works, in particular, the behavior of systems incorporating feedback was analyzed. But it is usual to consider 1948 as the beginning of control theory development, when the American scientist Norbert Wiener's *Cybernetics; or, Control and Communication in the Animal and the Machine* was published (see Wiener 1961).

Wiener was the first to show that a control principle incorporating the assistance of a feedback mechanism is common in technical systems and organisms. A general theory of systems (von Bertalanffy 1950), game theory (von Neumann and Morgenstern 1953), information theory (Shannon 1958), hierarchical systems theory (Mesarovic et al 1970), and optimal control systems theory (Bellman et al 1958; see Pontryagin 1990) began to develop at that time. Numerous works on information processing, the filtration of noisy signals, and adaptive behavior appeared (Ashby 1952). It is necessary to point out that even at that time, it was conceived that signal filtration could be achieved using predictive mechanisms, in which knowledge of the process that generates the signal to be filtered was used to create a model (Tou and Gonzalez 1974; Kalman and Bucy 1961). However, this mechanism was not widely used because the complexity of its technical requirements made it impracticable. As we will see later, nature, whose "technical abili-

Biological Neural Networks
Konstantin V. Baev
© 1998 Birkhäuser Boston

ties" surpass mankind's, used just such an idea for the construction of information processing systems such as the brain's.

The first automated calculating machines were built using a universal schema described in 1937 by the great English mathematician, Alan Turing, who proposed a theory of universal calculating automata, published at about that time (Turing 1936, 1950). From 1950, Turing was occupied with the mathematization of biological theory. He would undoubtedly have obtained remarkable results in this field, if he had not tragically died in 1954.

It is necessary to point out that in comparison with analog automatic control systems, the digital calculating automata have a different mathematical basis—so-called constructive mathematics. The algorithm, or effective calculating procedure, is a central notion in constructive mathematics. The algorithm usually consists of a set of descriptive instructions for converting input data into output. A finite number of elementary instructions, or commands, which could be performed by a universal automaton—Turing's machine—are used for making up the instruction set.

At that time, many scientists thought that the calculating abilities of Turing's machine were equivalent to those of the brain. This notion was based on Church's axiom, which states that an intuitive and informal determining class of algorithmical calculating functions corresponds to the class of functions being calculated by Turing's machine. It was noted, however, that unlike Turing's machine, the brain has the ability to perform creative work, e.g., independent problem formulation. So far, attempts to formalize these concepts have not been successful. Discussions about the issue still continue, but with a more philosophical than mathematical focus. In Chapter 13, we will explore the differences between the governing principles of Turing's machine and those of the brain.

The systematic development of ideas about modeling the thinking function with the help of automata constructed of networks of neuron-like elements was first presented in 1943, in the work of McCulloch and Pitts (1943). The main result of this work is proof that, given any algorithm beforehand, it is possible to build a network of neuron-like elements that will demonstrate the behavior described by the algorithm. At the end of the 1950s in the period of powerful cybernetic development, the works of Frank Rosenblatt, in which he described the *perceptron*—a neuron-like apparatus able to learn simple tasks of pattern recognition—were published. His main work, *Principles of Neurodynamics* (1962), was devoted to the investigation of the capabilities of these apparatuses. This book had a remarkable impact on the course of brain research and the modeling of brain function.

A separate developmental direction in cybernetics, known as "artificial intelligence," then appeared. The period had come when it seemed that the fundamental mysteries of brain function would soon be revealed, and an artificial yet comparable model of human intellect would appear. It was thought that this artificial intellect would probably even surpass human intellect, because the calculating speed of electronic machines had already surpassed that of man. These ideas still existed in the middle of the 1970s, but were reevaluated when it was realized that constructing such a model would be an immense undertaking, rather than an easy victory. New enthusiasm for the field of artificial intelligence emerged in the 1980s with the appearance of a new discipline—neurocomputing (see Section 6.1).

2.2 Basic Concepts of Control Theory

As mentioned in the Introduction, control theory has synthesized its own formalized language to describe the control process. Let us consider the basic concepts of control theory, in order to use them in future discussions. (More detailed information can be found in numerous textbooks on control theory [see, for example, Leigh 1987; Arbib 1995].) This analysis will also help us to understand the major hurdles that anyone attempting to create an automaton—a controlling device—must overcome. One will also see that nature itself encountered similar obstacles while developing biological neural networks during evolution.

Any act of control implies the existence of two interacting functional components: a *controlling device* and a *controlled object*. In general, one can formulate the concept of a control task for a controlling system in the following way. The task involves changing the state of a controlled object according to a specific rule or set of rules designated as a *control law*. This principle implies that, in the presence of rather complex control problems, the automatic controlling device has to be capable of (1) determining the current state of its controlled object by using information available from the controlled object, (2) storing (or acquiring from an external source) information about the state of the object at a given destination point, and (3) generating (computing) the necessary control influences on the object that permit the controlling device to achieve the control goal.

Physicists, mathematicians, and engineers have developed a specific language for describing the details of control processes. It is customary to use a system of coordinates to describe a given state of a controlled object. Such a system of coordinates is usually multidimensional, and uniquely describes the position of the object within a given space. This space is called a *state space*, and is a generalization of the term *phase space*, which is commonly used in physics to describe dynamic systems. In most of the literature, "phase space" is used synonymously with the more general "state space." Such a coordinate system differs from the more simplistic and commonly accepted concept of space coordinates. For instance, in the case of the unidirectional movement of a material body, phase space is two-dimensional—i.e., it is a plane, and the coordinate axes are position and its derivative, velocity. Quite often the success of an analysis of the behavior of a physical system depends on the type of coordinates chosen.

Therefore, the state of an object can be represented as a state vector, and a process of control can be formalized as a process of computation over components of an input state vector and generation of an output control vector. Obviously, it is possible to talk about a *control space* where absolute dimensionality is different from *state space* dimensionality. Such a description has the following major advantage: it considers control to be a process of motion in a state space such that a powerful mathematical apparatus developed for the analysis of physical dynamic systems can be applied to the analysis of control problems. It is noteworthy that a given state of a controlling system itself can be considered as a point in a definite state space, and the entire process, from start to finish, of *computation* can also be described as motion. The dimensions of state spaces vary from one system to another, and usually depend on the complexity of the particular system.

Let us try to imagine the major mathematical problems that are the consequence of such an analytical description. Figure 18 shows the relationship between a control device and a controlled object, which is widely known as control with feedback. At first sight, this scheme looks like a perfect control system. An input vector, consisting of information from sensors located on the object, is provided to the control device via feedback, and the control device transforms it into an output control vector. Such is the case for rather simple control tasks. In cases of more complex objects and their complex control systems, a designer will face many difficulties. Below, we will consider the most significant of these difficulties, and several illustrative neurobiological and technical examples will be utilized in order to facilitate a rational application of control theory to neurobiology.

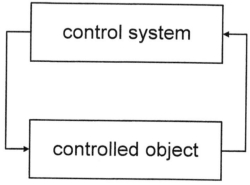

Figure 18—Schema of control with feedback.

The simplest of these difficulties is that information from sensors, representing the object state with some degree of fidelity, provides the substrate for an input vector, but this information usually requires preprocessing in order to produce the input vector itself. Some sensors can be nonlinear, some may duplicate each other, and others may have additional confounding features. The control system must therefore have a functional component that is capable of processing incoming information in order to produce a useful input vector for the control device.

Observability of a controlled object

The degree of observability of a given controlled object is high only for rather simple controlled objects. Complex objects usually have a very low degree of observability. Such is the case when there are many environmental influences on the object for which the control system cannot account. The latter situation eventually results in a diminished contextual appreciation of a controlled object whose observability suffers from contextual poverty because it includes only a part of the environment. When this occurs, additional sensors must be utilized for the object in order to encode a greater number of environmental parameters. This compensatory mechanism can help to solve the control problem, but it will obviously result in the construction of a more complicated control device, because of the necessity for performing additional computations over those additional parameters. Some of those additional parameters may be mutually dependent and, as mentioned earlier, may lead to additional computational problems. However, the most significant problems associated with *limited observability* arise because limited observability ultimately means that information received from the controlled object is noisy. Therefore, the control system has to possess additional computational abilities and has to spend more time to process this information and determine the current state of the object. Limited observability also means that the state of the object can be determined only within specific limits of probability.

Controllability of a controlled object

If a controlled object cannot be moved by a control system to a desired state in one control step, it means that the object has *a low level of controllability*. This is also typically the case for complex systems, and it obviously creates additional computational difficulties for the control system because a recursive computation of the control influence has to be performed in order to achieve a particular control goal. Later, we will discuss how important it is for a control system to possess a model of the controlled object in order to effectively control an object that has a low level of observability and controllability.

Optimization

The process of control can be conceptualized as the movement of the controlled object from an initial point to a destination point in its state space, while it travels along a discrete trajectory. The control vector for any given process cannot be chosen arbitrarily. There are always some limitations, but as a rule there exist a large number of possible controls. Therefore, it would be ideal to choose the control vector that is "the best" among all possible controls. This would mean that a given control satisfies a specific criterion. For example, if we need to move a physical body on a given plane from an initial point to a destination, we may choose the shortest distance between these points, from among all possible trajectories (Figure 19). The same would also be true for a multidimensional space, if an analog of distance were introduced. This is the major reason why a metric describing "distance" is usually defined for multidimensional state spaces. If it is necessary to minimize time consumption for moving a body from one place to another, the trajectory of such a movement does not always coincide with the shortest distance between the initial and destination points. Problems of this type are well known in the calculus of variations, and the first problem formulated by Johann Bernoulli in 1696, and referred to as the problem of

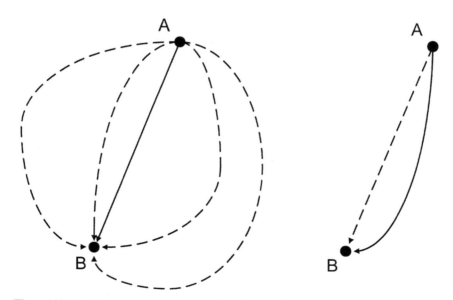

Figure 19–Possible movement trajectories from point A to point B. The shortest trajectory (left panel) and the fastest one (right panel). The fastest trajectory is the solution of the problem of brachystochrone—the curve of most rapid descent of a particle under the influence of gravity. See text for explanations.

brachystochrone, made a substantial impact on the development of this branch of mathematics. The problem consisted of determining the curve of most rapid descent of a particle under the influence of gravity. The solution of this problem showed that such a trajectory does not coincide with the shortest distance between the two points, as one might assume on the basis of intuition (Figure 19).

In technical fields, the solution—necessary control—is usually identified by finding a function that minimizes (or maximizes) a mathematical expression known as a cost function. This procedure is known as optimization, and it was developed by generalizing the theory of variations. Modern control theory is often referred to as optimal control theory because, as a rule, some control criteria are optimized. Obviously, different optimization criteria can be created, and unique control solutions can be obtained in this way, if the corresponding mathematical equations that describe the control process are solved. It is customary to use the term "global optimum" to refer to a situation in which all possible criteria for the control process have been optimized. It is worth mentioning that the criteria describing reliability, costs, control system structure, etc., can be among the criteria that are optimized in the setting of a global optimum. A formalization of these problems leads to the necessity of finding extrema of functions for corresponding cost-functions. For the purpose of the discussion that follows, it is necessary to mention that in control theory optimized criteria are usually subdivided into two groups—the first group describes the structure of the system and the second its parameters. A vector of parameters can include such components as weights, delays, amplification coefficients, time constants, etc.

A designer must have a mathematical model of the control process in order to create a reasonable controlling device, and the model has to be as precise as possible. Such a model mimics the behavior of a natural process, provides information about state and control spaces, and ultimately determines the structure of the controlling device. However, the term "model" can have a variety of other meanings. It is possible to speak of a model of a desired result, a forward model, an inverse model, etc. The model of a result implies that a control system compares a desired model state with the real state of the object, and a mismatch signal is used to compute the necessary control. A typical example of the use of a model of a desired result is temperature control in a refrigerator. A desired temperature is compared with the real one, and a cooling system turns on or off depending on the result of this comparison. In the majority of practical applications, a model of the result is used. This is the major reason that the term *model* is most often associated with a model of a result.

A *forward model* mimics the causal flow of a process by predicting its next state (for example, the position and velocity of the object), given its current state and the control command. *Inverse models* invert the causal flow by estimating the control command that causes a particular state transition. The terms *forward model* and *inverse model* are broadly used in motor physiology. However, there is a strict definition in control theory of what motor physiology refers to as a forward model. It is a *model of controlled object behavior*, and this term will be used below.

It is well known in optimal control theory that any optimal control system has to use *a priori* information in the process of its control. One of the most important uses of *a priori* information is for defining the current state of a controlled object that has a low level of controllability and observability, and whose future state can therefore only be predicted with a limited degree of probability. *A priori* knowledge is usually present in the form of conditional probabilities for certain events. Everybody who has studied the theory of probability knows Bayes' formula, which ascribes *a priori* knowledge to some events. Similar themes are utilized in control theory, and a number of algorithms have been developed to use *a priori* knowledge. One example of this is the so-called filter of Kalman-Bucy. Additional examples of the application of *a priori* knowledge are well known in communications theory, in which it has been demonstrated that it is much easier to identify a signal in a noisy channel if the shape of the signal or its time of arrival is known. However, technical applications of such algorithms to control processes are fraught with many difficulties, and much simpler models, such as a model of a desired result, are most often used, as mentioned above.

If one summarizes the history of the development of control theory, one will find in the end that it is a history of the development of mathematical methods for analyzing control processes, i.e., the history of the computational methods of control theory. In the discourse that follows, we will use the most universal of interpretations of the term "computation" which encompasses an approximation of various mathematical functions as well as the development of their corresponding computational algorithms.

Two principal approaches for performing computations are utilized in control theory: digital computation and analog computation. The first is based on so-called *constructive mathematics* and is strongly dependent on the level of sophistication of the hardware and software being used. The fact that digital devices can be programmed gives them much more flexibility than analog devices. In complex control systems, both digital and analog control devices must have specific memory capabilities—short- and long-term memory—because, for example, they have to store both information about the control law and the results of any intermediate computations. Short-term memory can be built on the basis of short-lasting processes such as transitory processes that have discrete time constants. Examples of long-term memory are well known to any computer user, and are exemplified by the many different types of disks, tapes, etc. In the case of analog computation, long-term memory is based on structural hard-wiring and on the properties of elements included within the corresponding structural schema.

Programming, however, can be quite complex and time-consuming, and not all computational problems can undergo easy algorithmization. This is why in the past two decades, neurocomputing—a new field that uses a network-based computational principle—has become so popular for solving computational problems for control needs. We will discuss this approach in Sections 2.3 and 6.1.

2.3 Computational Abilities of Biological Neural Networks

If one considers the nervous system to be a controlling device, then it is necessary to know something about its computational abilities. Presently, we know little, almost nothing, about the computational abilities of biological neural networks. None of the simulations that were born exclusively within neurobiology and dedicated to the modeling of electrical properties of single cells and neural networks provides us with this information. However, there are several mathematical abstractions that can give us a clue to how powerful the computational abilities of biological neural networks can be. In 1943, Warren McCulloch and Walter Pitts published a famous paper that was based on the computations that could be performed by a two-state neuron. They probably were responsible for the first serious attempt to understand what the nervous system might actually be doing. Such neural computing elements were mathematical abstractions of physiological properties of actual neurons and their connections.

The *McCulloch-Pitts neuron* is a binary device, i.e., it can be in only one of two possible states. The neuron has a fixed threshold and receives inputs from excitatory synapses that have identical weights. Excitatory influences combine linearly, and a time quantum for integration of synaptic inputs is based loosely on the physiologically observed synaptic delay. The neuron can also receive inputs from inhibitory synapses. Their action is absolute, meaning that if the inhibitory synapse is active, then the neuron cannot become active. Obviously, such a neuron is capable of performing so-called *threshold logic*.

The central conclusion of that paper is that any finite logical expression can be realized by a network built of McCulloch-Pitts neurons. It was a very important conclusion, because it showed that rather simple elements connected in a network could have immense computational power, and it also suggested that the brain is a powerful logic and computational device.

Another mathematical abstraction, Kolmogorov's theorem, reveals a network computational principle and shows which functional forms can be approximated by neural networks. Kolmogorov's theorem provided the first clear insight into the versatility of neural networks for use in function approximation. It appeared in 1957 and astounded mathematicians. The history of this theorem may be found in an article by Vera Kurkova (1995).

The proof of the theorem will not be presented here. It can be found in the original citation (Kolmogorov 1957). This theorem states that a mapping network consisting of three layers of processing elements can precisely implement any continuous mapping function. Kolmogorov's mapping neural network existence theorem is presented below in the form used in neurocomputing (see Hecht-Nielsen 1990):

Given any continuous function $f:[0,1]^n \to R^m, f(\mathbf{x})=\mathbf{y}, f$ can be implemented exactly by a three-layer feedforward network having n fanout processing elements in the first (input) layer, $\mathbf{x}$; $(2n+1)$ processing elements in the middle layer; and m processing elements in the top (output) layer, $\mathbf{y}$.

The theoretical construction of Kolmogorov's neural network is shown in Figure 20. The processing n elements of the first (bottom) layer are fanout units that simply distribute the input $\mathbf{x}$ vector components to the processing elements of the second, hidden, layer. The $2n + 1$ processing elements of this hidden layer neither directly receive inputs from the outside world nor provide outputs directly to the outside world. The transfer function of these units is similar to that of a linear weighted sum. The m output processing elements of the third (top) layer send signals to the outside world, the output $\mathbf{y}$ vector. The transfer function of output units is highly nonlinear. This is strictly an existence theorem and tells us that such a three-layer mapping

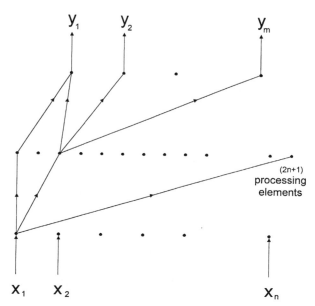

Figure 20—Kolmogorov's network. The first layer consists of input fanout units. The second and third layers have processing elements with semilinear and highly nonlinear transfer functions, respectively. Explanations are given in the text.

network must exist. This theorem is not constructive, because it does not show us how to build such a network.

Since Kolmogorov's description, others have demonstrated both theoretically and practically that the so-called *backpropagation* neural networks can implement a function to meet any practical need by using a learning procedure (see Hecht-Nielsen 1990). Different backpropagation learning laws have been proposed. Furthermore, there are many other learning laws that are used to train neural networks (see Chapter 6).

Theoretical analysis of the computational abilities of artificial neural networks has convincingly shown that many mapping functions can be stored within a network, and that the quality of the performed approximations degrades rather slowly with the increase in the number of stored functions. Therefore, inherent in the very nature of the network computational principle is a remarkable capacity to be multifunctional. Given this information alone, one can draw a very important intermediate conclusion: The mechanistic approach analyzed in Chapter 1 is not applicable to such multifunctional computational systems.

On the basis of Kolmogorov's theorem, it is also possible to gain an insight into the versatility of a real neuron for use in function approximation. Most real neurons have an immense dendritic tree whose size significantly surpasses the size of the neuron itself. It has been accepted for a long time that synaptic "buttons" located on dendrites are capable of generating only local synaptic potentials, and that these potentials spread electrotonically to the neuronal soma, where summation of the synaptic influences occurs. If the result is higher than the threshold, then the neuron generates an action potential. Within such a framework, action potentials are not generated in dendrites, and it is hard to imagine that a synapse located far from the soma, for instance at a distance of several millimeters, can produce a substantial electrotonic influence on the neuronal soma, given the cable properties of the dendrites. Another point of view describes dendritic spike generation that significantly expands neuronal computational capabilities. The possibility of spike generation in dendrites has been experimentally demonstrated in several laboratories, and it has been demonstrated that the extent of dendritic spike propagation can be effectively modulated by synaptic potentials (see Buzsaki et al 1996; Tsubokawa and Ross 1996).

Figure 21 demonstrates how it is possible to consider a single neuron in terms of Kolmogorov's theorem. Suppose a dendritic spike moves towards the soma and reaches a dendritic bifurcation. There are two possibilities. The first is that it crosses the bifurcation and continues moving towards the soma because the spike can evoke a superthreshold potential in a wider region of the dendrite. The second is that the spike evokes only a subthreshold potential in a wider region of the dendrite, and therefore its propagation is stopped. Obviously, in the second case the superthreshold excitation of a wider dendritic region is possible when two spikes in thinner dendrites occur

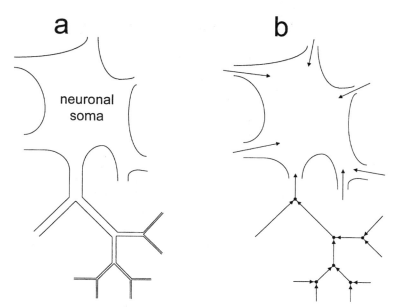

Figure 21—An example of the application of Kolmogorov's theorem to a neuronal dendritic tree. Left panel, schematic drawing of an actual neuron with a part of its dendritic tree. Right panel, presentation of the part of the dendritic tree shown in the left panel in the form of Kolmogorov's network. Each dendritic bifurcation is considered as a nonlinear node of Kolmogorov's network. Arrows designate directions of dendritic spike propagation. See text for explanations.

simultaneously, or when an excitatory synaptic potential occurs in a wider dendritic region when the spike arrives. Therefore, it is possible to consider a dendritic bifurcation as a nonlinear node of Kolmogorov's network. It is also clear that the reality is much more complex, and various other mechanisms are possible because dendrites are covered with numerous excitatory and inhibitory synaptic buttons, and any spatial combination of them is possible. Moreover, dendrites of some cells have specific formations, such as spines, and they can also create dendrodendritic synapses.

Even the simplest combinations, such as those described above, permit us to consider a neuron with a part of its dendritic tree as a network that reminds us of Kolmogorov's network with a single output (Figure 21). In the case of a real neuron, the number of layers can be significantly more than three. Different branches of the dendritic tree can have different functional meanings; they can participate in computations of different functions.

Therefore, it is clear theoretically that even the unidirectional neural networks utilized in neurocomputing possess great computational abilities. As a matter of fact, such networks perform only recoding of input to output.

Such a network is only a reactive system. It responds with an output only to an input, and the response is a function of the external input combined with specific network properties, e.g., the configuration of the synaptic weights and properties of neurons.

Real neural networks, as a rule, are not unidirectional and have numerous negative and positive feedback loops. Thus the question appears: how does this influence the calculating power of biological neural networks? Only the most general answer to this question has been obtained in neurocomputing and computational neuroscience (see Hecht-Nielsen 1990; Churchland and Sejnowski 1992). Obviously, in such networks, the effect of information received from an external input is not isolated from the information that arrived in the net just before it, and any network output depends on both external and internal inputs. Any possible network response to inputs varies between two polar situations: (1) The influence of internal inputs is small, and the net always has an output when there is an external signal. In this case, network output is modulated by internal input; and (2) an internal input alone is capable of activating an output, while an external signal concomitantly modulates output.

Feedback or recurrent networks also possess several important capacities. Among other capabilities, they can generate oscillations of varying durations, process temporally extended input sequences, incorporate multiple time scales into the network units, and resolve ambiguities.

As previously shown, the presence of feedback loops in real neural networks can also be interpreted as a necessary condition for both (1) the improvement of afferent information processing (because they provide a substrate for the function of an internal model within the system (see below)), and (2) computation of the output of a controlling system. The presence of feedback loops increases the calculating ability of the network, because feedback loops make it possible to use recursion in situations when the function value for a given argument value can only be computed using its previous values. Recursive computation is very important in many cases—for example, when it is impossible to lead the object to a desired state during one control step. In some networks, recursive computation can continue until a desired result, satisfying a specific calculation criterion, is reached. It is well known from the mathematical theory of algorithms that the class of recursive functions has maximum functional power. Therefore, from a computational perspective real neural networks have much more functional power than the unidirectional networks mentioned above.

Unfortunately, this is the only conclusion that one can make on the basis of the available data. Several problems related to the computational abilities of biological neural networks remain unsolved. Presently, we know little, almost nothing, about the computational abilities of single neurons and neural networks that include pacemaker neurons, circuit triggers, complex synapses, etc. Obtaining this information will have a profound influence on the issue of interrelations between function and structure (see Conclusion).

In the discourse to follow, the foundation of any new conceptual theme will be based on the conclusions made above. We will accept, without providing mathematical proof, that at each controlling level all functional subunits included in the network have sufficient computational abilities to control specific physiological functions. However, the existence of complex life forms with advanced nervous systems can be considered as a proof of the highly sophisticated computational abilities of biological neural networks.

Lately, the term *computation* has become more and more broadly utilized in neuroscience. However, one need only engage oneself in discussions with neurobiologists before one realizes that many scientists are still not willing to accept the use of this term in neuroscience, because they believe that it is neither necessary nor sufficient to describe the operations performed by biological neural networks (a typical biogenic point of view). Their major objection consists of the following: It is human beings who *perform computations*. Biological neural networks do not. They believe that neurobiology can still successfully describe the function of biological neural networks by using such terms as "signal transformation and processing," "reflexes," "excitation," "inhibition," and other cherished biological concepts.

Obviously this point of view is very anthropocentric, but there is still room within it for the idea that biological neural networks can perform computations, particularly at the highest levels of the human brain. Otherwise, human beings could not perform mathematical computations. In his review article "Neurons, dynamics and computation," published in *Physics Today* in 1994, John Hopfield presents the advantages of using the term *computation* to describe the function of biological neural networks. The tacit message in this review is that the complete acceptance of this term in neurobiology has not yet been achieved. It would be to the advantage of the neurosciences as a whole, if those who do not appreciate the potential of this term to further advance the neurosciences would accept the fact that this term will permit neurobiologists to apply the power of mathematics to describe the function of biological neural networks.

Computational neuroscience is most often characterized as an approach to understanding the information content of neural signals by means of modeling the nervous system on many different structural levels (including the biophysical, circuit, and systems levels). Computer simulations of neurons and neural networks are considered complementary to traditional techniques in neuroscience.

Unfortunately, one has to admit that the utility of mathematics is presently limited because it can only be applied to the analysis of the simplest of neural networks, and much still has to be done before mathematicians can begin describing networks with the complexity of biological neural networks. However, even current mathematical approaches are already much more constructive than the biological ones. We also do not yet know the most effective way to approach the problem of mathematical descriptions of the functions of biological neural networks (see Conclusion). Does this mean that until this problem is solved we do not have a hope of adequately describing the mechanisms underlying the highest of brain functions and drawing fruitful conclusions? The answer might be yes if we believe that the only avenue leading to our understanding of the complexity of brain function is the accumulation of more and more detailed knowledge of neuronal properties and their interconnections, concluding with a final model of the network. This, as we have already seen, appears to be the dominating current strategy. The answer might be no if we start by providing a functional description of the system and revealing its organizational principles. In this case, an understanding of the computational capabilities of specific networks would follow from the conceptual and functional descriptions. We would be limited primarily by the rather simple modeling constraints required to demonstrate the principles of a network capable of performing the specific type of computation necessary to result in a given function of the nervous system. The second alternative is the avenue advocated in this monograph.

The constructive meaning of the term computation for neurobiology will be set out in the discussion to follow. However, one additional useful term, *automatism*, has to be introduced in order to make our discussion even more constructive.

2.4 Broadening the Concept of Automatism: Inborn Automatisms and Acquired Habits

Let us go back to our proposal that the nervous system is a controlling system created by nature. Given the data accumulated by modern neurobiology, it is possible to conclude that the capacity for automatic control is the principal feature of the nervous system. At birth, an animal possesses a vast variety of inborn automatisms, and it acquires new ones during its lifetime. Moreover, the higher an animal is on the evolutionary scale, the deeper is the significance of its spectrum of inborn and acquired automatisms. Disabling an animal by lesioning different brain regions abolishes the automatisms controlled by those regions. As a rule, the most evolutionarily recent areas of the brain control more complex automatisms. For example, a lesion in the highest of brain levels (e.g., cortex) in mammals abolishes not only previously acquired automatisms, but the ability to produce new ones as well. At the same time, inborn automatisms remain intact.

The reader will have noted that we have extended, to some degree, the concept of an automatism as compared to the definition that is usually used in neurobiology to describe inborn behavior. The definition now includes acquired skills. In our everyday life, we do usually define acquired skills are automatisms. The essence and the necessity of such a broad generalization of the concept of automatism will become clear as one reads through this book's pages. It should be noted here only that a more comprehensive understanding of automatisms enables us to analyze the function of the brain from a unified perspective. Reflexes and program control in combination with feedback control are automatisms. The logical conclusion of all this is that *learning* can now be considered to be the process of forming new automatisms.

In the light of this new appreciation for the utility of a more generalized definition of an automatism, the major questions facing modern neurobiology must now be reformulated. Instead of traditional questions that focus on the construction of different reflex arcs within the nervous system and on neural networks that generate programs for effector organs, the more fruitful and universally applicable question should be introduced: what is the basis for the automatisms of a nervous system and its different parts? As will be shown later, the answer to this question will form the basis of our ability to overcome the crisis in modern neurobiology and to formulate a more comprehensive theory of brain function.

The relationship between the terms "automatism" and "computation" is quite similar to the relationship between a goal and the means by which it is achieved. In general, to create an automatism, it is necessary that the appropriate adjustments be made in the computational abilities of each functional subdivision within a particular control system.

2.5 How Can Control Theory Concepts Be Applied to Biological Neural Networks?

One can and should ask oneself the above question after reading through the previous text. The answer to this question is rather transparent. This theory is applicable to biological neural networks if the way in which they function is "optimal." But what will it mean for neurobiology if the answer to the question, "Do biological neural networks display optimal function?" is yes? How does one prove that the function of biological neural networks is optimal? From where must one start? Should we prove it for all nervous systems and their different levels? If it is proven for the highest neuronal level, will it or should it mean that all lower levels are also optimal in their function? These and other questions come naturally to mind. Let us try to find answers to these questions. The best way would be to start from a discussion of the term "optimization."

In our everyday life, we use this term to designate something that is better than the other things in its class. Of course, this use implies that some criterion or criteria were applied to make a comparison. We can talk about the optimal *solution,* the optimal *construction,* the optimal *shape,* etc. Most often, biologists and other specialists in nontechnical fields use this term in this context alone; it does not possess any additional constructively applicable meaning.

In control theory, as we saw, the optimization process has a powerful constructive meaning, and the strategy of synthesis—in which investigation of function provides the basis for any subsequent formulation of structural description, and our understanding proceeds in this manner—can be proposed as a more fruitful and appropriate alternative to the currently dominant *analytical* strategy in neurobiology. This approach is not new to engineers. They usually create technical controlling systems in this manner, from a given base of elements necessary to solve a proposed problem. In this regard, any control algorithm can be realized using as its basis very simple elements such as neurons (see Section 2.3). It is also well known that the same information processing application can be implemented by different means: mechanical devices, electromechanical relays, electronic circuitries, optical devices, chemical reactions, etc. The difference among such devices is the speed at which they perform computations, and hence their computational speed determines their effectiveness. As a rule, an improved base of system elements significantly simplifies the schematic manifestation of any given control problem.

Let us imagine a situation in which the same task of creating a control system is given to several engineering groups who have the same base of elements. Moreover, the primary requirement is, for example, to minimize material waste. It is well known that in such situations the different groups will usually make approximately identical schematic decisions. Identical decisions are also observed when other requirements—for example, to maximize work speed, minimize energy waste, etc.—are given.

Therefore, if, for the creation of a controlling system, there is a set of elements with pre-determined features, and there are criteria of optimization that embrace all salient parameters of the system in question, then there can be only *one* resulting schematic decision (a so-called *global optimum*) that appropriately and comprehensively satisfies these criteria. On the other hand, optimization will lead to different decisions if different sets of elements are used to create control systems with the same function. Obviously, if this idea is applicable to biological controlling systems, then we have only to focus upon a functional description of these systems, because there can only be one optimal structure of a such a system. Here the reader has a right to remark: "We as people create systems in such a way, but is there any reason to believe that nature creates neuronal controlling systems in the same way? And if there is, then how did it perform optimization?" All subsequent sections are devoted to answering these questions.

If one begins to think in this way, one will at once realize the vague nature of the term *function*, and the limitations imposed by our contemporary understanding of the term as it has historically been used in physiology. The descriptive terms that follow from this incomplete understanding of *function* are familiar to us: function of the muscle, cerebellum, spinal cord, neural center, feedback, ion channel, etc. Such an application of the term *function* is a natural consequence of the current dominant analytical approach used in physiology. It is clear that such a limited understanding of function is not appropriate or suitable for the function-structure approach that is possible for control theory. The concept of function must therefore have a systems meaning in our approach, and, what is more important, is that it must be constructive. Let us return to innate motor automatisms, for example. Now it is typical that a CPG is involved in producing a motor program. But the generator—for instance the locomotor generator—is only a part of a control system that also contains muscles and peripheral feedback mechanisms. Control of rhythmic limb movements is the function of this entire system. In turn, this system is a subsystem of the whole neuronal system of the brain. The isolated consideration of a single portion of a complex system, and the substitution of functional investigation for the analysis of the external display or expression of the work performed by this isolated portion of the entire system, does not permit the establishment of functional criteria for the system's optimality. Nor can it reveal the organizationally complex nature of the neuronal connections, without which we cannot understand the structure of the system.

Naturally, this broad comprehensive understanding of function requires a change of methodology in brain research, because the methodology we use determines the way to perform our analysis of the components of a complex control system. First, the system being investigated must satisfy the requirement of "closedness." This calls for the investigation of the processes taking place in all of the system's components: controlled object—controlling brain region and interconnections between these processes. When a concrete function is performed in a closed system, it is possible to establish the relationships among controlling signals, controlled object behavior, afferent signals, and the results of their processing in the nerve center. Such an approach is the exception and not the rule in contemporary neurobiology. Furthermore, any experiments performed on this system must be directed toward obtaining information about the *function* of the system. In other words, each experimental fact must be interpreted in the context of the overall function of the whole control system being investigated. Obtaining experimental data without such interpretive consideration may lead to a significant waste of time and energy, as well as to an accumulation of useless facts.

The combination of experimental findings with theoretical descriptions is a necessary condition for success in utilizing this newly proposed strategy for neurobiological investigations. Only by applying this strategy can an adequate description of the function being investigated be provided. As the previous section demonstrates, serious mathematical application can provide the basis for theoretical investigation. Applying functional descriptions only in terms of physiological processes, as is traditional in contemporary neurobiology, is not sufficiently systematic and constructive for advancing the science of neurobiology. As was already mentioned in the Introduction, the successful application of the proposed strategy described in this monograph to some concrete neural control systems could quite possibly lead to the appearance of some universal theory of brain function. Such a theory would have to be constructively applicable by the scientist to concrete functional tasks performed by any part of the brain. In turn, after the creation of such a theory, its applicability to other controlling systems regulating the behavior of different object types would have to be demonstrated. The applicability of the theory to other non-neuronal biological control systems will be shown in the Conclusion.

For the optimal combining of theoretical and experimental approaches to solve this lofty task, it is necessary to choose, from the very beginning, a neuronal subsystem that is simple enough to allow for theoretical analysis and, at the same time, complex enough to produce results of fundamental interest. As seen in Chapter 1, the systems controlling innate automatic movements satisfy these requirements.

The perspective of the neurobiologist on the CPG problem was described in previous sections. Let us try to imagine the mathematician's point of view on this problem. For most biologists, it has been unclear (yet important) until now why long-term rhythmic processes with time constants of hundreds of milliseconds or even seconds are possible in neural networks consisting of individual elements that can integrate input signals with short time constants consisting of a few milliseconds. Mathematicians do not see any problem with this. It is easy for them to show that quite long processes can appear even in a network with as few as three neuron-like elements. Moreover, the theorem of McCulloch and Pitts solves the problem of rhythm generation because rhythm can appear in the network without having any specific mechanisms.

Afferent flow is usually presented in generator models as tonic initiating signals. It appears that scientists reject the necessity of a theoretical solution to the problem of generator interaction with phasic afferent flow. Some scientists think that afferent feedback is not important for the work of simple generators (see, for instance, Arshavsky et al 1985a) because rhythm generation is observed in its absence. But for the mathematical specialist in automatic control theory, it is the problem of generator interaction with afferent flow that is precisely the most interesting.

The discourse above, in this section, permits one to draw the following conclusion. If it is possible to prove that the controlling system responsible for an inborn motor automatism such as locomotion or scratching, is an optimal control system, then we can generalize an optimality principle and conclude that the higher levels of control are also optimal. Obviously, this generalization will be more persuasive and impressive than the opposite situation. The proof that the highest motor control levels are optimal will not necessarily mean that the level of control involved in regulation of inborn automatic behaviors is also optimal. Here, we have come back to a question that is similar to the one mentioned in the beginning of this section: How can we prove that a controlling system that satisfies the requirement of "closedness" and includes a CPG is optimal? The best proof would be to show that any CPG possesses a model of object behavior. From a theoretical point of view, the presence of such a model is a necessity for any CPG, because the controllability and observability of its controlled object (for example, a limb) are rather low, and hence, if the controlling system is optimal, it has to have a model of object behavior. Other proofs such as the proof that a controlled object moves along an optimal trajectory, that the control influence is optimal, etc., are complementary but appear much more time-consuming and less persuasive.

Let us again go back to the use in the existing dominant control theory paradigm of the term *model*. It is based on the classical schema of control using the mismatch between the model of result and feedback information (see Figure 22). It is described in all books of automatic control theory and can be found in the neurobiological literature (Bernstein 1966). It has been suggested that the neural center produces both the efferent program and the model of the result of an action, the acceptor of action as per Anokhin (1974). Sometimes this is referred to as predicted sensory "reafferences." The neural center uses the *efference copy* (von Holst and Mittelstaedt 1950) in order to create the model. Internal rings—intracentral reverberations of excitation caused by the relaying of the output signals to the input of the neural center—are considered to be the basis for such interaction. The concept of the model of the result is very similar to the notion of feedback in cybernetics. Afferent information coming from the effector organ is compared with the model, and any resultant mismatch is used for program correction. However, no one has ever carried out an experimental investigation of the existence of peripheral correction in rhythmic motor generators based on this concept of the function of a model in control theory. Until now, this suggestion has been the domain of theorists, but not of experimenters.

It is necessary to stress that Figure 22 shows only the model of a result, not the functional model of the controlled object. In other words, the result shown depicts the necessary final state of the controlled object. This schema is simple and convenient for solving such tasks as homeostasis regulation,

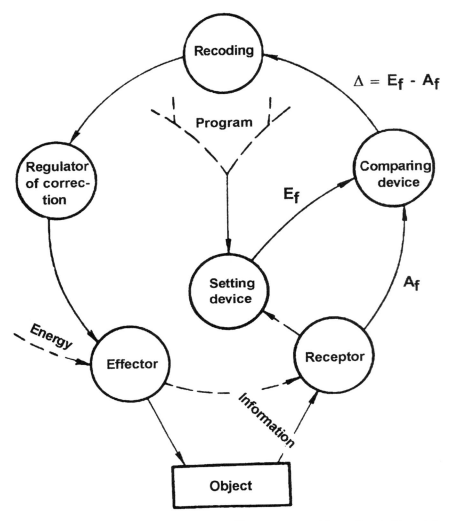

Figure 22—Schema of a movement controlling apparatus. (According to Bernstein 1966). E_f and A_f, expected and actual values of parameters, respectively; $\triangle$, mismatch value.

but it does not give a system the ability to perform tasks after the feedback has been cut off. The model result must be dynamic during cyclic movement control, defining the state of the controlled object for each moment in time. The parameters of its dynamic state must be the same as those of the controlled object. But even in the case of total coincidence of these parameters, the working of this system will be disrupted after cutting the feedback.

Undoubtedly, the ability of the spinal cord and invertebrate ganglia to control "correctly" any rhythmic movements after partial or total deafferentation contradicts such classical schemata. It will be shown in Chapters 3 and 4 that afferent information processing is based on another principle. As a result, the functional schema of generator interaction with afferent flow is the darkest part of the "black box" known as the "central pattern generator." Obviously, that is why the topic of rhythm generation has come into the scientific limelight. As we have seen, the method of searching "under a street lamp" is not the best in this case. All paths leading to an understanding of the generator pass through the solution of the problem of generator interaction with peripheral sensory flow.

It is worth mentioning that in human motor physiology, the concept of a model is quite broadly used. However, such a model is usually a model of the result (Wolpert et al 1995), and it has not yet been proven experimentally that the highest levels involved in motor control in human beings possess a model of object behavior. Nowadays it is possible to find theoretical postulates in which the basic principle of interaction between peripheral and model signals consists of their comparison and subtraction. For example, motion sickness is considered a result of sensory conflict (Oman 1988) following comparison of model and real flows.

Therefore, the importance of combining program and feedback controls in the same controlling system remains unappreciated. The importance of this property provides the foundation that makes it possible to propose hypothesis for an experiment to prove the existence of a model in CPGs. The ability to cut peripheral feedback can be used for the investigation of regularities in the changes occurring within the generator motor program under the influence of an experimentally determined afferent inflow, and it is also quite easy to analyze afferent signals coming to the generator from the moving limb during actual movement. For all that, the rationale might be the following. There are definite relationships between the phase sensitivity of a generator to afferent signals and the distribution of those signals in the motor cycle. If such relationships are ascertained, these facts will be the key for understanding the laws governing the construction of the whole system. Besides the abovementioned relationships, the physiological essence of the facts obtained through those experimental models will become known, because the system meets the requirement of "closedness."

The investigation of reorganization in the generator program under the influence of afferent signals probably reminds the mathematician of the method of "black box" identification, which is well known in cybernetics. In other words, for a mathematician, such a novel neurobiological approach resembles an attempt to investigate any system created by nature from the standpoint of control theory. While understanding the attractiveness of this approach, we still must be conscious of the fact that success depends on the correct choice of experimental models. Invertebrate generators (as the reader has seen) are built in the most simple way, and correspondingly, afferent information processing also must be simple. Generators are significantly more complex in vertebrates. One must suppress the initial desire to perform such investigations on invertebrates or on lower vertebrates, because the "simplicity" of the experimental object would inevitably lead to obtaining only "simple" knowledge. This is why it is better to choose an experimental paradigm with an object whose complexity is high enough to preserve the chance for success of the experiment. From the point of view of control theory, this means making the choice of an optimal investigation strategy. With these considerations in mind, it is better to choose the cat as our subject. Many basic data of modern neurobiology were obtained from the cat. Two experimental models of automatic movements—locomotion and scratching—are widely used on cats. It is quite easy to investigate on a macroscopic level all of the substrates of the controlling system in the feline, as well as the kinematics of feline movement itself. To this end, it should be stated that the ability to observe different motor programs produced by the same part of the nervous system is necessary for revealing the common principles of their control.

3

A Central Pattern Generator Includes
A Model of Controlled Object:
An Experimental Proof

The data presented below were obtained from fictitious and real scratching and locomotion experiments. During actual movement, the integral characteristics of afferent inflow were investigated. In fictitious movement, the quantitative analysis of rearrangements in the motor pattern under the influence of tonic afferent inflow (as conditioned by ipsilateral hind-limb position) and phasic afferent inflow (simulated by stimulation of hind-limb afferents in different motor cycle phases) was carried out.

For the quantitative analysis of generator efferent activity, the duration of the entire cycle and different cycle phases, as well as corresponding root mean square values of efferent discharges, was examined. Mean square measurements (later the term "intensity" will be used) were selected because they reflect the effective value of efferent discharges (by analogy with electric current), i.e., the square was proportional to the power of these discharges. The same divisions of the cycle of movement as mentioned in Section 1.1.3 were used in these experiments: the scratch cycle was divided into aiming and scratching jerk phases, and the locomotor cycle was divided into flexion and extension phases. These terms will be written in quotation marks when designating their corresponding phases during fictitious movements.

Systematic investigation of the relationships between stimulation factors (limb position, stimulated afferent input, stimulus intensity, and stimulation phase) and corresponding motor pattern rearrangements has led to the collection of an enormous volume of data. A complete description can be found in previous papers (Baev and Esipenko 1988a, b; Baev et al 1991a, b; Baev and Shimansky 1992; Esipenko 1987; Shimansky 1987; Shimansky and Baev 1986, 1987a, b). These experimental data will be examined briefly in this section. The major experimental facts and corresponding conclusions are as follows:

Biological Neural Networks
Konstantin V. Baev
© 1998 Birkhäuser Boston

1. Scratching and locomotor generators, when deprived of their tonic afferent inflow, produce motor programs that are in good accordance with the physical laws of corresponding rhythmic motions.

When tonic afferent flow influences the generator during fictitious movement, its motor program is most similar to programs present after deafferentation when the hind limb is at the physiologically natural position for the scratch reflex (aimed) or for locomotion (lowered). In other words, when the afferent feedback is cut off, the scratching generator "implies" that the hind limb is aimed, and the locomotor generator implies that the hind limb is lowered. The functional role of motor program reorganization mechanisms driven by tonic peripheral afferent flow is thus to provide stable aiming or lowering of the hind limb.

2. During actual scratching and locomotion, bursts of afferent activity coincide with transitions between cycle phases.

Schemata summarizing the results of the investigation of integral afferent activity in lumbosacral dorsal roots during scratching and locomotion are shown in Figures 23 and 24. Two components—tonic and phasic—were observed in both cases. Periodic fluctuations in afferent activity were observed against a background of tonic increase (not shown in the Figures). The peaks of afferent activity bursts coincided with the transitions between efferent activity phases.

3. Stimulation of afferent inputs during specific phases of fictitious movement does not change the pattern of motor activity. These phases were designated as no-rearrangement (NR) points.

To simplify the description of this regularity, the changes in cycle duration will be the primary consideration. Despite the difference in the cycle duration reordering plots, corresponding to different afferent inputs at different stimulus intensities and positions of the ipsilateral hind limb, certain general features were distinguished (Figure 25):

- At least one part of the reordering plot (RP) has a positive slope intersecting the level corresponding to the absence of (or zero) duration change.

- NR points can be found exclusively within discrete regions of the stimulation phase. These regions are unique because transitions between cycle phases occur within them. Some NR points present during locomotion coincided with transitions between phases on the contralateral side (Figure 26).

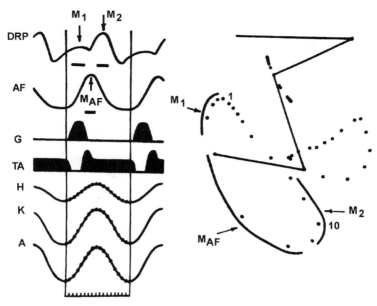

Figure 23—Relation among changes in dorsal root potentials, integral afferent flow, joint angles, and limb movement trajectory during scratching. Left panel, normalized changes in dorsal root potential (DRP), integral afferent flow, and joint angles. Right panel, movement trajectory of limb tip calculated on the basis of joint angles. Interval between dots, 20 ms. AF, afferent flow; G and TA, electromyograms of m. gastrocnemius and m. tibialis anterior, respectively; H, K, and A, angles in hip, knee, and ankle joints; M_{AF}, M_1, and M_2, maxima of afferent flow, first and second components of dorsal root potentials. Solid lines designate maxima most frequently observed.

- The amplitude of reordering, and consequently of the steepness of the reordering plot, increases with a rise in stimulation intensity; the steepness of the negative slopes of these plots rises sharply with increasing stimulation intensity until they turn into breaks between the positive slopes (Figures 25, 26). The uppermost (left) point of the negative slope remains virtually static, and this part shortens at the expense of its lowermost (right) point (or the starting point of the following positive slope), shifting to the left.

The dependence of cycle duration changes on hind-limb position consisted of a rather sharp rise in the amplitude of reordering produced by a stimulus that coincided with a discrete position of the hind limb. At this position the RPs were similar to those produced by high stimulation intensity. During stimulation of cutaneous afferents, such an effect was observed when the hind limb was deflected backwards (Figure 25). In the case of

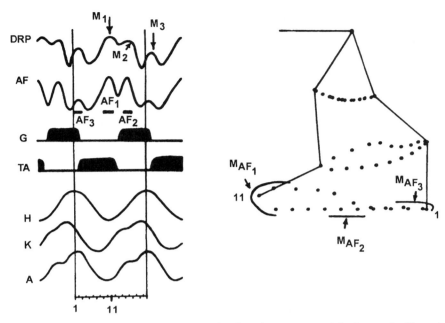

Figure 24—Relations among changes in dorsal root potentials, integral afferent flow, joint angles, and limb movement trajectory during locomotion. AF₁, AF₂, and AF₃, first–third maxima of afferent flow, respectively. Other designations are as in Fig. 23.

muscle nerve stimulation, the rise in reordering amplitude was observed at the limb position that induced the stretching of muscle agonists (deflecting the hind-limb backwards stretched the aiming muscles, and over-aiming it stretched the scratching muscles). An increase in reordering amplitude with the strengthening of phasic stimulation of limb afferents, or the post-addition of tonic afferent inflow to the phasic stimulus at the corresponding hind-limb position, seems quite natural. A stronger stimulus indicates a greater movement disturbance for the generator, and it has to take "drastic measures" to correct the motor pattern.

The dependence of cycle duration changes on afferent input was expressed in the correspondence between inputs and their specific NR points.

Because of the dependence of intensity changes on stimulation phase, afferent input, stimulation intensity, and limb position had, as a rule, a complex character; a detailed description will not be provided here. It is important, however, to note two main peculiarities. The phases of zero change coincided with NR points (of cycle duration) in most but not all cases. The tendency of reordering amplitude to rise in response to stimulation increase was also observed in changes of motor activity intensity. It was also possible to

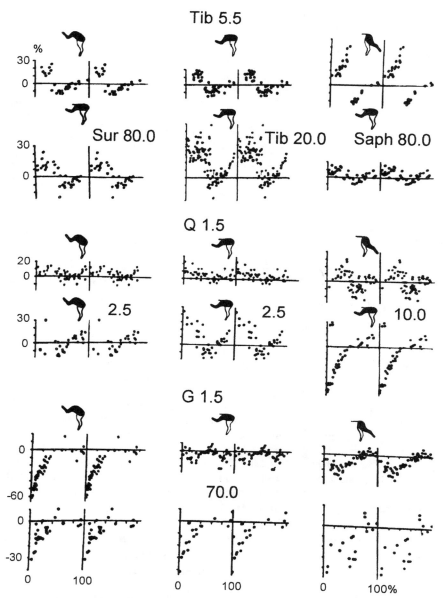

Figure 25—Relationship between phase of stimulation (abscissa) and corresponding changes in "scratching" cycle duration (ordinate) at different hind limb positions. Tib, Sur, Saph, Q, and G, n. tibialis, n. suralis, n. saphenus, n. quadriceps, and n. gastrocnemius, respectively. Stimulus intensity in denominations several times above threshold of the most excitable fibers in shown in numerals. Pictograms show hind-limb positions ("over-aimed," "aimed," and "deflected backwards") from left to right in, e.g., the upper row of the scatter diagrams; the detailed description of these positions is given in Baev et al 1991a.

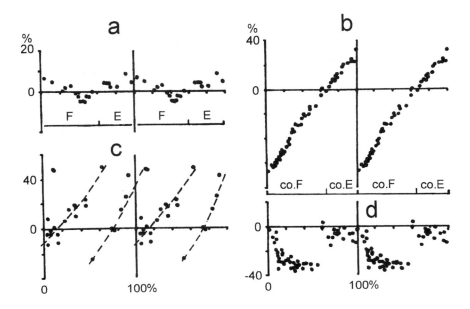

Figure 26—Relationship between stimulation phase (abscissa) of the cutaneous tibial nerve and corresponding changes in duration of "locomotor" cycle (ordinate). a, low-intensity stimulation during "locomotor" rhythm with frequency of 0.5–1.0 Hz; b and d, plots for the cycle where stimulus was applied, and for subsequent cycle, respectively. High-intensity stimulation during "locomotion" of the same frequency as in (a); c, high-intensity stimulation during "locomotor" rhythm with frequency of 1.5–2.5 Hz. Division of "locomotor" cycle into "flexor" (F) and "extensor" (E) phases is shown below the plots. Cluster analysis showed that in case c the no-reordering (NR) points clearly corresponded to different cycle duration ranges (the left and right NR points to "trot" and "gallop," respectively). In addition, both the NR points corresponded to the contralateral transition from "flexion" to "extension."

note the reversal of intensity changes after increasing the intensity of nerve stimulation. Thus, one may suppose that changes in cycle duration and in the intensity of motor activity are mediated by different neural mechanisms.

4. The existence of NR points gives the generator the ability to synchronize its activity with periodic afferent bursts.

Let us consider the positive slope regions tracing the zero change level obtained on the reordering plots of cycle duration, because their presence testified to the existence of a mechanism providing stability of synchronization between the "motor" rhythm and the periodic sensory signal, with the stable phase of the signal in the "motor" cycle being an NR point. This mechanism works as follows: it assumes that synchronization has set in, and that the phase of sensory signal arrival coincides with the NR point. Suppose that as the result of an error in generator function, the synchronization is disturbed, and the efferent activity starts to outstrip the afferent signal (i.e., the signal phase shifts to the right along the RP). Then the current cycle will lengthen, and thereby the outstripping of the generator rhythm will be compensated for. As a result, the phase divergence between it and the afferent signal will be compensated for. If the generator activity begins to lag behind the sensory signal (i.e., the signal phase shifts to the left along the RP), then the current cycle will be shortened and thus compensate for the delay. Thus, the phasic relationship between generator activity and the afferent signal—when the latter phase matches the NR point—is stable. The steeper the positive slope region is, the stronger the negative feedback stabilizing phase relation between generator activity and afferent inflow. It is apparently this mechanism that provides the generator with the ability to synchronize its activity with sensory signals (the phenomenon of entraining).

For example, when stimulating hind-limb nerves at a frequency that approximates fictitious scratching, synchronization of these two processes was observed. Distinct peaks on the histogram corresponded to NR points (Figure 27). The steeper the RP of cycle duration in the neighborhood of the NR point was, the clearer was the peak expressed.

In the light of the above finding, the mechanism of the synchronization of the generator activity with afferent inflow becomes clear. During actual movements, afferent activity bursts coincide with zero reordering phases (Figures 23, 24). Hence, it is possible to draw a more general conclusion from the information summarized in subsections 1–4:

If a burst of activity is observed in a particular afferent input at some phase of ongoing undisturbed rhythmic movement, stimulation of this input during fictitious movement at the same phase does not change the pattern of motor activity. The major goal of generator activity reorganization under phasic peripheral afferent influence is to bring the motor program into a dynamic relationship with sensory inflow. This causes minimal changes in generator motor activity, compared to what transpires after limb deafferentation. In other words, such a dynamic relationship is stable; this is true not only for the sum of different afferent influences, but also for any separate type of afferent flow. The aimed hind-limb position for scratching and the lowered position for locomotion could be considered as NR-points in the state space of various limb positions.

The initial notions of a "correct" movement trajectory and the internal model of the controlled object

If tonic afferent flow activates the mechanisms providing stability to the hind-limb position, what then is the physiological nature of the dependence of motor activity reorganization on the stimulation phase? To answer this ques-

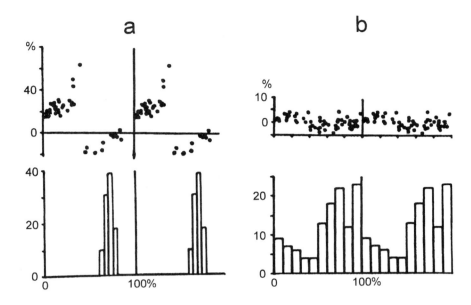

Figure 27—Correlation between cycle duration reordering plots and the ability of the generator to synchronize its activity with periodic afferent bursts. Upper plots are cycle duration reordering plots (constructed using the results of stimulation at random phase of "scratching" cycle); lower plots are the histograms of stimuli application at different phases with regular stimulation of the same nerves. a and b, stimulation of nerves subserving sartorius and gastrocnemius muscles, respectively. See text for explanation.

tion, let us assume that there is no rhythmic fluctuation of the limb, no phase modulation of the activity changes under afferent influence, and that the generator must simply support a certain position of the limb. When the limb deflects from this position, the generator will rearrange its activity to restore the initial position. Such a mechanism can evidently prevent fluctuation of the limb. It appears that the coordination of this mechanism with limb rhythmic movement can only be achieved by making it phase-dependent so that the mechanism cannot "notice" deflection of the limb during a certain phase interval. Thus, owing primarily to the property of phase dependence of generator response to a sensory signal, the generator can stabilize not only a discrete position of the limb but also the trajectory of its movement. The parameters of this trajectory must then obviously determine the pattern of the phase dependence. Following the principle of rearranging minimization, as described in the previous paragraph, one can conclude that the trajectory corresponding to the rearrangement minimum must be equal to that corresponding to the generator motor activity after limb deafferentation. Such a trajectory can therefore be designated as the "correct" trajectory.

During movement of the limb along such a "correct" trajectory, the generator receives correspondingly organized afferent inflow, which may also be referred to as "correct." According to the principle of afferent influence minimization, the generator does not react to such "correct" afferent inflow, i.e., the generator activity is not changed under the influence of this inflow. Changes in the generator motor program will take place only in cases in which afferent inflow becomes "incorrect" under the influence of external factors disturbing the movement. Therefore, the generator must have the model of expected afferent inflow corresponding to each motor program, if it is to obey the principle of minimal interaction.

The notion of an internal model evokes several questions. Its functioning must obviously be connected with the tonic and rhythmic activity of certain interneurons. Which ones? Is it possible to distinguish them from other types of generator interneurons? Does the model work during preserved afferentation in the intact animal, and if it does, how do the model and peripheral inflows interact? It is possible to answer these questions by taking into account the fact that primary afferent depolarization (PAD) is evoked both by generator and peripheral afferent signals.

5. The maximum of centrally-elicited PAD (or the maximum of negativity of DRP [dorsal root potential]) coincides with the concentration area of NR points.

In light of the abovementioned hypothesis, it is logical to propose that centrally-modulated PAD reflects the continuous production of model sensory activity in the generator. Bursts of model sensory inflow should correspond to NR points because generator activity is "correct" in the absence of peripheral afferent inflow. Thus, maximum PAD (thus the maximum of negativity of DRP) evoked by the generator must encompass the region of concentration of NR points. Such an association is shown in Figure 28 for the example of scratching. It is also observed for locomotion (Baev et al 1991b).

A careful analysis of the oscillations of cord dorsum potentials (CDP) during generator activity deprived of peripheral feedback offers one additional (practically direct) piece of evidence for the existence of model afferent inflow. When CDP oscillations are aligned side by side with the wave corresponding to dorsal root potential (DRP), it is possible to observe bursts whose phase corresponds to the moments of transition between cycle phases (Figure 29), i.e., to NR points. Obviously, these bursts reflect the activity of dorsal horn interneurons, i.e., of the interneurons receiving peripheral afferent signals. Experiments in which rhythmically and tonically active interneurons receiving afferent inputs were identified in the spinal cord during fictitious scratching and locomotion (Baev et al 1991a, b; see also Section

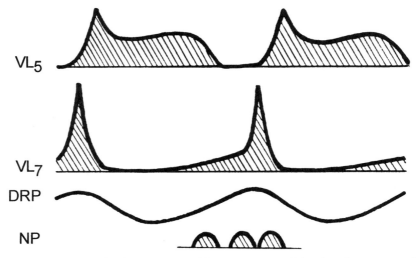

Figure 28—Correlation between scratching generator activity, dorsal root potential (DRP), and distribution of NR points (NP) in the "scratching" cycle. See the text.

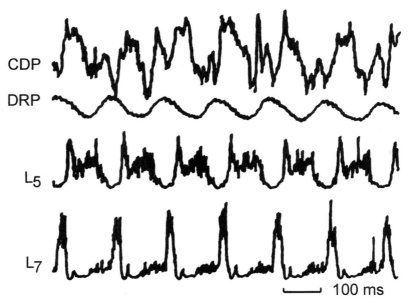

Figure 29—Cord dorsum potential (CDP) oscillations and their relation to oscillations of dorsal root potential (DRP) during fictitious scratching. L_5 and L_7, efferent activity in the corresponding spinal cord segments.

1.1.3) provide evidence of the presence of interneurons involved in rhythm generation and afferent information processing. These interneurons are activated by the generator itself just at the moment when afferent inflow bursts are observed during movement of the limb along a "correct" trajectory.

6. Model sensory inflow is produced during real movement with intact feedback.

The pattern of DRP during actual scratching and the relationship between it and integral peripheral afferent activity are shown in Figure 30 (see also Figure 23). The two main phasic components of DRP are distinguished within them. Experiments have shown that the first and second components are, in general, of central and peripheral origin, respectively, with the first one coinciding with NR point concentration. These data give clear evidence that the internal sensory flow is produced when peripheral feedback is intact as well. Phasic malalignment between "internal" and "peripheral" PAD components apparently occurs when the movement trajectory is not optimal, such as when the limb does not brush against (misses) the body surface to be scratched. It is necessary to stress that malalignment allows the scratching generator to "see" what the hind limb misses. The total afferent activity determined by the deafferented hind-limb movement in accordance with this undisturbed motor program becomes phasically aligned with centrally originated PAD (Figure 30d-f). This fact can hardly be explained without a hypothesis that includes the sensory model. Obviously, the same rationale can be comprehensively applied to a locomotor generator that also modulates PAD (Baev 1980; Baev and Kostyuk 1982).

From the point of view of the above-stated concepts, the interaction between peripheral and model afferent flows is the basis for peripheral correction. This takes place not only at the postsynaptic level, but also at the presynaptic level. Model flow changes the peripheral signal influences on central neurons as the result of central PAD modulation. It is not unreasonable to believe that peripheral flow, in turn, can change model flow influences due to presynaptic mechanisms. The latter may be possible, if there are axo-axonal synapses on the terminals of spinal neurons, as have been demonstrated for several types of existing spinal cord neurons (Maxwell and Koerber 1986). These questions will be considered in detail below.

7. Presynaptic inhibition of central origin modulates reactions to peripheral afferent stimulation.

It is possible to clarify the way in which afferent signal inhibition is connected with the time course of DRP evoked centrally, by observing changes in responses to peripheral stimuli applied at different phases of the motor generator cycle. It is convenient to analyze the amplitude of the N_1-compo-

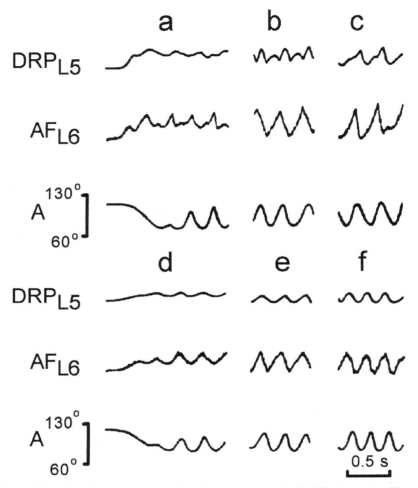

Figure 30—Correlation between dorsal root potential (DRP), integral afferent inflow (AF), and ankle joint angle (A) during scratching. a, initial aiming stage; b and c, stable scratching; d–f, the same as a–c, respectively, but after deafferentation of the hind limb.

nent of CDP evoked by stimulation of cutaneous afferents, because nonsegmental interneurons take part in its formation (Bernhard and Widen 1953; Lindblom and Ottosson 1953). Their activity is virtually unrelated phasically to the rhythm of the generator. Results of similar experiments have been described earlier (Baev 1981a, b). It was shown that the maximum of DRP negativity corresponded to the lower amplitude of the N_1-component, and the minimum of DRP negativity to the higher one.

Results of a more detailed investigation during fictitious scratching are shown in Figure 31. It reveals that the amplitude of the N_1-component is inversely proportional to the amplitude of DRP, i.e., to the value of afferent terminal depolarization (Figure 31a). Moreover, the maximal response amplitude is lower than the one present prior to generator activation. The decrease, obviously, corresponds to the tonic component of centrally evoked PAD.

How does presynaptic inhibition evoked by model flow modulate the action of the peripheral afferent signals that effectively reorder the generator rhythm? It is possible to reveal such modulation by investigating the steepness of the reordering curves for cycle duration. As is shown in Figure 31b, the curve steepness for "scratch" cycle duration is inversely proportional to the value of DRP centrally evoked (compare with Figure 31a). It was difficult to demonstrate a similar regularity for a locomotor generator using the curves of its cycle changes, because there was a much bigger (in comparison to "scratching") scatter of dots in the diagrams, evoked by greater fluctuations of the "locomotor" cycle.

Therefore, both weak and strong afferent signals are modulated. This reflects the difference between post- and presynaptic inhibition. The first could

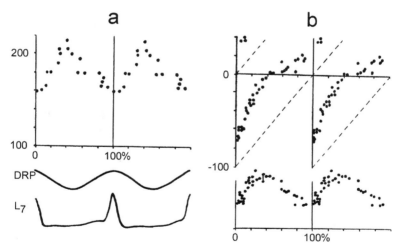

Figure 31—Central modulation of peripheral influences on the scratching generator. a, dependence of the N_1-component amplitude on the phase of stimulation in the "scratching" cycle. Dorsal root potential (DRP) and activity in the L_7 ventral root are shown below for comparison. It is clear that N_1-component amplitude is inversely proportional to the DRP amplitude. b, dependence of the steepness of the "scratching" cycle reordering plot on the phase of stimulation (n. quadriceps, strength of stimulation equaled 10 thresholds). Upper diagram, reordering plot. On the lower diagram, each point ordinate is equal to the distance along the ordinate between the dotted lines and the corresponding point in the upper diagram. Explanations are given in the text.

be interpreted as an increase in the excitation threshold, as a result of which a weak signal coming to a postsynaptic neuron does not evoke an action potential. Presynaptic inhibition is, to a certain degree, equivalent to a synaptic weight decrease between a depolarized terminal and a postsynaptic neuron.

8. DRPs of central and peripheral origin interact with each other on a parity basis.

The amplitude of the evoked DRP depends on the size of central PAD in a manner similar to the dependence of the N_1-component (Baev 1981a, b). This leads one to think that centrally evoked presynaptic inhibition weakens the influence of the peripheral afferent signal not only on neurons of reflex arcs, but also on neurons generating PAD. The question then arises: will there be a weakening of central influences on the PAD generating system during the development of PAD evoked by peripheral afferent flow? The answer to this question has been demonstrated to be positive. The peak of DRP negativity during its modulation by the scratching generator is significantly reduced when the stimulus is applied before the development of this negativity (Figure 32). It was supposed earlier that it is necessary to consider centrally evoked DRP as a result of activation of the PAD generating system by model flow. From this point of view, the above-stated data indicate that peripheral and model afferent flows inhibit each other using the PAD generating system.

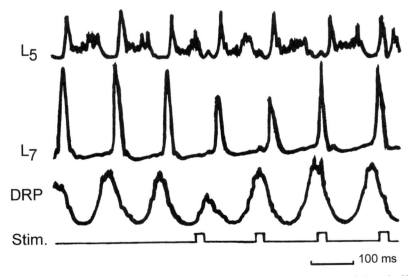

Figure 32—Inhibition of centrally evoked dorsal root potential by peripheral afferent signals during fictitious scratching. The afferent signal was simulated by stimulation of n. quadriceps with the strength 2 thresholds.

Therefore, it is possible to conclude that *the interaction between peripheral and afferent flow occurs according to the laws that describe the interaction of different components of peripheral afferent flow*.

Such a concept enables one to resolve what might appear to be contradictions in the interpretation of experimental data regarding changes in the tonic PAD component during activation of spinal generators for scratching and locomotion. According to one interpretation, there is an increase in the tonic PAD component during scratching generator activation, and a decrease during activation of the locomotor generator, i.e., hyperpolarization of central afferent terminals (Baev 1979, 1980; Baev and Kostyuk 1981, 1982). These data explain the changes in different segmental reactions during activation of both generators as opposed to the state of "rest" (Baev 1979, 1980). Other researchers have obtained data that demonstrate an increase in the tonic PAD component during activation of the locomotor generator (Duenas and Rudomin 1988). Our experimental results also show that there can be an increase in PAD during activation of the locomotor generator evoked by stimulating the midbrain locomotor region, and that there is a decrease in DRP amplitude evoked by peripheral afferent stimulation during intense fictitious locomotion (Figure 33).

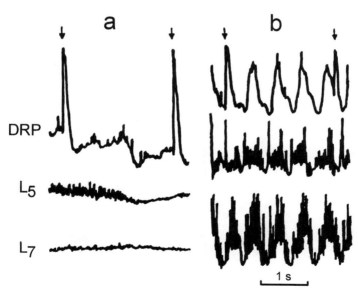

Figure 33—Decrease of the dorsal root potential (DRP) wave evoked by peripheral nerve stimulation during an increase in locomotor generator activity. An animal was immobilized (fictitious locomotion). a and b, "locomotion" of low and high intensity, respectively ("gallop" in the second case). Moments of afferent stimulation (n. tibialis, strength of stimulation was 3 thresholds) are shown by arrows.

From the concepts developed here, it can be seen that the tonic PAD component registered at "rest" is induced by peripheral tonic afferent flow, i.e., by limb position, while the tonic component observed during generator activation is a result of a complex interaction of peripheral and model flows. The appearance of model afferent flow exerts a double-directed influence on the PAD level: on one hand it additionally activates the PAD generating system, and on the other it weakens the influence of peripheral afferent flow upon it. Thus, the resulting change in PAD level depends on the relationship between these interactions. This relationship in turn depends on various features that determine the general state of the spinal cord, and, in particular, on the substrates of the PAD generating system. So far, the character of these features is unclear. At the very least, the decrease in PAD level in comparison to the "rest" state (Figure 34a) might be observed during activation of the scratching generator (in an immobilized animal). The presence of intense antidromic discharges prior to generator activation is an indication of increased activity of the PAD generating system. DRPs evoked by peripheral nerve stimulation decrease after activation of the scratching generator (Figure 34b).

Therefore, the facts described above provide substantial evidence that can be considered to be proof of the existence of a putative internal sensory model within generators of rhythmic movements. Mutual influences between the inner analog of sensory flow embodied within the model and peripheral sensory flow emanating from sensory organs are symmetrical. Later, such interactions will be referred to as "parity interactions." The functional essence of the presence of a model and its interaction with peripheral sensory flow will be discussed in detail in the next chapter.

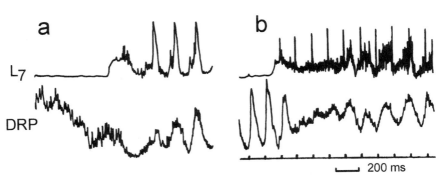

Figure 34—Integral primary afferent depolarization (PAD) changes during activation of the scratching generator. a, inhibition of PAD background activity; b, suppression of DRPs evoked by peripheral nerve stimulation (low-threshold stimulation of L7 dorsal root). Periods of stimulation are shown below.

4

The Spinal Motor Optimal Control System

It was previously noted that the function of biological systems would best be described from both functional and substrate perspectives. Let us begin the analysis from a functional point of view, so that we may reveal what the system must "be able to do." We will then move on to the description of the substrates of biological systems with regard to neural networks, in order to consider how an actual system consisting of neurons that have specific functional properties might be constructed. Since this text is intended for biologists and physicians, there will be no mathematical formulas in this discourse. However, those who wish to explore some of the mathematics that provide the basis for the discourse to follow can find them in the Appendix (see also Baev and Shimansky 1992). The corresponding references to the Appendix are provided in the text to follow.

The essence of the solution of the "black box" problem rests within the postulate that control procedures for both locomotor and scratch movements should be considered to be different regimes of a general (spinal) optimal motor control system (Baev and Shimansky 1992). From this viewpoint, the existence of an internal model of controlled object dynamics is a consequence of optimality of the control system. This model should be considered an additional internal "sensor" that complements other limb sensors, with the goal of supplying the control system with information about the most probable behavior of the controlled object for any given moment in time. After limb deafferentation, the control system relies solely upon the information present within the model. If the model information is adequate, and no significant accidental perturbation occurs, the control system will produce the proper control action. Thus, we should call "the generation of a rhythmic motor pattern" a reverberation, not between the generator half-centers of flexion and extension, but rather within the loop of internal feedback that transfers model "sensory" signals (see Figure 35).

Biological Neural Networks
Konstantin V. Baev
© 1998 Birkhäuser Boston

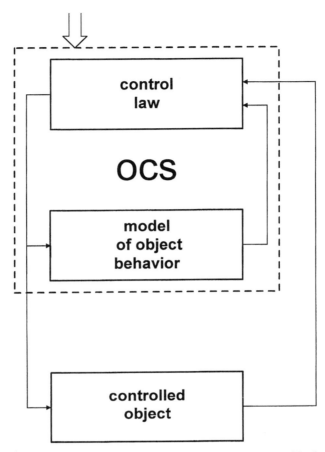

Figure 35—General schema of the spinal optimal control system. Explanations are given in the text.

Let us go back to control theory. According to the mathematical concept of an optimal control system, a given control value is a function (control law) of the current state of the controlled object. In the case of the spinal motor optimal control system, the afferent signals carry information about the current state of the controlled object, the subsystem processing sensory information recognizes this state, and then a decision is made about what control action should be produced for the current situation. In other words, the optimal control system (OCS) receives from the controlled object the information about the current values of its state space coordinates, and then, according to the optimal control law, converts these coordinates into a control action that moves the object in its state space along an optimal trajectory, which is based upon specific criteria, to the aim state. If there is a description of the behavior of some object in the form of a differential equation system, then the phase space dimension is equal to the number of initial conditions necessary for solving the system of equations. In practice, phase space coordinates correspond to the number of variables of a differential equation system. Let us consider these ideas in the example of scratching movement control. We will regard a limb as a mechanical system that is a mass point with certain forces applied to it. These are the forces developed during active contraction of muscles, the forces conditioned by the viscosity and elasticity of the muscle-tendon system, the gravitational force, etc. In this case, the minimum set of phase space coordinates may consist of the usual space coordinates of the limb tip and velocity components. The simplest phase space is therefore six-dimensional. The aim state set by an initiating signal may then be determined by three space coordinates of the irritant location and three components of the limb tip velocity vector in the region of the irritant. The velocity should be sufficient to scratch the irritant out. It is natural to regard the optimality criterion in this case as minimization of energy and time expenditures.

The problem surrounding the criterion of control system optimality seems significant. The complexity of the nervous system presumes a multicriterion character. It is possible to put forward the criterion of minimum simplicity of the OCS neural structure as additional to the criterion of minimization of energy-time expenditures. The problem of the relationship between the different criteria—which of them is the major one—can be solved in principle on the basis of experimental data only. The experimental results permit one to think that, in the case of scratching, energy-time minimization is the principal and most decisive criterion. In other words, a neural control system, at least in vertebrates, is organized according to the principle "as it is necessary to be optimal" rather than "as it is possible." The importance of this principle for the development of any novel theory of the function of the nervous system should *never* be overlooked.

The above-stated viewpoint rests on several assumptions:

First, the structure of a neural network controlling a given body part is fully determined by the fact that the anatomy of this part is optimum for its function. Here a highly constructive principle is influencing the system, and on this basis any given value of any parameter of the control system can be determined as optimal. Such an approach to the construction of a control system model differs in principle from that which is generally used for phenomenological modeling, because the latter is focused not on the optimality of the function performed by the system, but on the superficial similarity between the model and the behavior of the system.

Second, both the postural and the cyclic movements may be considered to be different regimes of the general optimal control system. The type of movement is obviously determined solely by the pattern of initiating signals. If the aim phase state has nonzero velocity components, the movement will have a cyclic character; otherwise, it will be of the postural type (see Figure 36). In other words, a certain movement can be programmed by determining the goal of control (we will return to this issue below).

Third, the nature of the phase dependence of so-called correctional reflexes (i.e., changes in the generator motor program in response to stimulation of peripheral afferents) can be clearly understood. Such a stimulus carries information about the current phase state of the controlled object, and the corresponding reaction of the control system should depend on the intrinsic conclusion drawn by the control system regarding the current phase state. If these states appear equal, the stimulus "tells" nothing new to the generator and, therefore, does not evoke any change in the motor pattern. As was shown in the previous section, the stimulation phases in which no changes are produced (NR-points) have been clearly observed experimentally.

Let us consider in detail the nature of the so-called "afferent correction" term used in neurophysiology. A schematic of limb phase trajectory during scratching is shown in Figure 37. Any afferent signal entering the scratching generator will be regarded as information about the phase space of the limb as a controlled object (point S') at the current moment. In actuality, a certain difference will be observed between this state and the one "calculated" by the internal model of the controlled object (point S). This mismatch may be naturally divided into two components: phase (Δ_φ) and

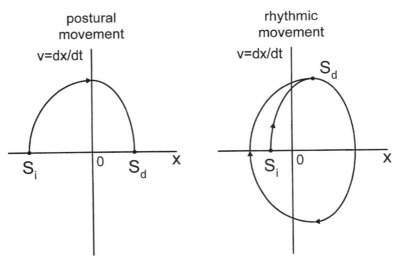

Figure 36—Schematic interpretation of postural and rhythmic movement trajectories in a phase space. S_i and S_d are the initial state of the controlled object and the state of its destination, respectively. See text for additional explanations.

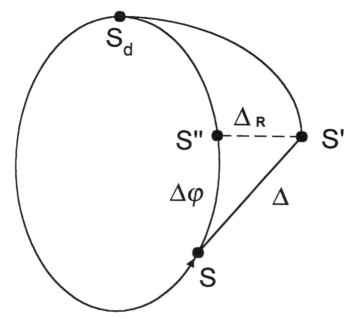

Figure 37—Division of limb deviation (Δ) from "correct" trajectory into phase (Δ_φ) and radial (Δ_R) components. See text for explanations.

radial (Δ_R) (Figure 37). The first phase reflects time delay between the current real and model states of the limb, and the second radial one reflects the limb deflection from the optimal trajectory. The phase component can be compensated for by the corresponding change in scratch cycle phase, without significant expenditures of time and energy. To compensate for the radial component, it is necessary to change the relationship between the activities of the flexor and extensor muscles.

It was noted in Chapter 3 that the absence of changes in cycle duration does not match the absence of changes in motor activity intensity in all cases. Now we can give a simple explanation for this fact. The mismatch between the real and the model states of the controlled object may have a zero value for the phase component, and a nonzero value for the radial component. Such a situation takes place when we apply high-intensity stimulation to cutaneous afferents during fictitious locomotion (Baev et al 1991b). Thus an absence of change in both cycle phase and motor intensity can be achieved simultaneously only if there is a certain relationship between stimulation phase and the intensity of an afferent activity burst. Experimental proof of this relationship may be the reversal of changes in motor activity intensity (noted in Chapter 3) during fictitious scratching after a transition from middle- to high-threshold stimulation of cutaneous afferents.

If there is a nonzero value for the phase component of the mismatch, an instantaneous (in comparison with cycle duration) change in motor rhythm is observed (Baev et al 1991a, b). It is important that the value of motor pattern reordering depends on stimulus intensity. A decrease in the latter leads to a widening of the segment of the reordering plot that has a negative slope, where the value of phase change has a definite random character with a rather large dispersion. These effects reflect a complicated process of interaction between peripheral and model information flows.

According to the concepts developed above, model "sensory" inflow interacts with peripheral flow just as isolated components of peripheral inflow interact with its other components. Despite the fact that the abovementioned schema allows a principal explanation for the ability of the CPG to work correctly without the aid of peripheral feedback, it does not answer the question of why partial deafferentation does not lead to systematic error when the system is processing sensory information. Thus the main attention should be focused on how the mechanism of interaction between different sensory inflow components takes place. To determine its organization, it is necessary to analyze in detail the features of the whole sensory information processing subsystem in the spinal cord.

4.1 Sensory Information Processing in the Spinal Motor Control System

Here and below, sensory information processing is considered the transforming of sensory inflow into some internal representation of the current state of the controlled object.

The interaction between the OCS and the object controlled by it can be described in the language of mathematical game theory, utilizing special types of games: so-called "Games with Nature." From this point of view, the controlled object's behavior should be regarded as the realization of its strategy in the game with the OCS. It is necessary for the OCS to select its strategy in the form of control law, so as to minimize the value of its "loss" as determined by an optimality criterion. It is clear that the optimum choice of strategy in any game can only be performed when an opposite strategy is properly recognized. In game theory, it is customary to use such notions as "loss" or "payments." If the information about the controlled object's behavior that is received by the OCS allows the determination of at least the probabilities of its different strategies, the OCS can determine a mathematical expectation of "loss" value for using a particular strategy.

The optimum strategy of control should obviously be selected according to the principle of minimizing the value of loss. If the strategy of the controlled object's behavior is not recognized properly, the OCS will "lose" a certain additional value. From a mathematical point of view, the optimum estimate of the object state (strategy) can be found from the condition of extra payment minimization. It is clear that any additional payment reaches its minimum when the probability estimate of the object state is maximum. It may appear in a general sense that any extra payment for an error in strategy recognition sharply increases for a particular strategy of controlled object behavior. Then, if the probability that this critical strategy will be used in the current moment in time is not equal to zero, it may be more advantageous for the OCS to select its strategy according to the critical strategy, despite the fact that the probabilities that the controlled object will use other strategies are significantly higher. However, in most real situations any extra payment depends rather weakly on object strategy, and the strategy of control should consequently be chosen according to the most probable strategy of controlled object behavior.

Thus it follows from motor control system optimality that the system should seek to obtain the maximum amount of reliable information about the current phase state of the controlled object, calculating the current control value, in the simplest case, according to the most probable object state.

The physical and chemical features associated with transmitting electrical activity in a nerve fiber determine how information is encoded. One can distinguish the following parameters of a neural signal: the duration and the amplitude of electrical impulses (spikes), the interspike interval (or its inverse value—spike frequency), and the location of signal origin within the neural network. The smallest parameter variations are observed in the duration and in the amplitude of spikes. In certain cases (which will be considered below), however, the amplitude undergoes significant changes that have a functional load.

The peculiarities of encoding signal value by spike frequency are rather specific. An external observer utilizing this model of a neural network can set any dependence of spike frequency on time, and thereby consider the signal value to be known (to the observer) precisely in every time moment. However, it is clear that a device receiving impulses cannot, in principle, determine the frequency of impulses instantaneously if the device does not contain an exact model of the impulse source. But if it does, the receiving device is able to predict the arrival of the next impulse precisely. So the conditional entropy of the source in the receiver and the quantity of receiving information are equal to zero. In the case of model absence, the frequency of impulses can be determined only approximately, and then only during a nonzero time period. On the other hand, when the value of a rapidly altering signal is encoded by impulse frequency, there is a limitation to the maximum frequency of signal alteration, which can be no more than half of the impulse frequency. These considerations force one to introduce the notion of *reliability* of information received by the nervous system. Obviously, if a control system varies its attention to different afferent channels depending on their reliability, then it can determine the current object state with a high accuracy by using a set of comparably imprecise measuring devices.

The projective field of limb afferents includes a very large number of sensory information channels. The excess of incoming information is tremendous. There is a practical and simple duplication of information in many channels (e.g., muscle afferents of the same type). Signals passing through afferents of different types and/or from different parts of the limb correlate strongly with each other during limb movement and result in a spatial surplus of information available to the network. A limb as a mechanical system possesses inertia and comparative invariability of various parameters, which in turn result in strong correlations between sensory signals entering the spinal cord at different moments in time, producing a temporal surplus of information. Despite the approximateness of information in every single channel, the spatiotemporal surplus of information inflow from a limb that is a controlled object gives the spinal OCS the potential ability to achieve high accuracy in determining the current phase state of the object. It is clear that this ability can only be realized if the OCS has a sufficiently complex and powerful sensory information processing subsystem within it.

Every separate receptor of the limb is a rough measuring device. Moreover, the reliability of information transferred through an individual afferent fiber deteriorates gradually with decreasing impulse frequency encoding the intensity of receptor stimulation. If the frequency is, e.g., 50 Hz (which in reality is quite typical), the duration of an interspike interval is 20 msec or 10% of a typical scratching cycle duration. Such liberal time allowances for frequency encoding cannot permit a precise determination of the current controlled object state. Thus, the lower the frequency of discharges, the less reliable the information transferred.

The presence of an attention mechanism in the spinal cord is well known. It is presynaptic inhibition. It can permit, for instance, the aquisition of a weighted sum of incoming signals by paying the most attention to the most active channels. Let us consider the simplest example in order to demonstrate how this feature can be realized in the nervous system. Suppose a neuron has two inputs that are functionally inhibiting each other by utilizing this presynaptic inhibition mechanism (Figure 38). If each channel contributes one conventional unit of information, and it is reduced to the level of 0.5 by functional cross inhibition, then the total sum is 1 (Figure 38a). If one of the channels is broken (has a zero value), then the sum is still one, because there is no inhibitory influence from the broken channel (Figure 38b).

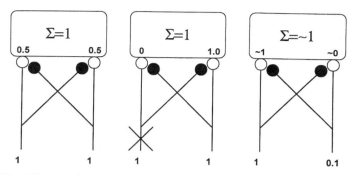

Figure 38—The simplest example of how "attention" to a more intensive channel can change the result of a weighted sum.

Finally, if one of the two channels has low intensity, then the sum is still close to one (Figure 38c).

As has been shown in Chapter 1, the preservation of adequate functional capabilities after partial disruption of information channels is peculiar to the brain. This property has been referred to as "holographic," in analogy to a hologram, in which every individual part contains an image of an entire object, but with decreased resolution power. Now, given the above, we can conclude that *a neural sensory processing system possessing holographic properties considers the absence of impulses from an information channel (afferent fiber) to have zero informational reliability (full unreliability), but not to have reliable information about the zero value of a sensory signal.*

From a formal point of view, we can say that a mathematical function determined over a variable number of arguments possesses a holographic property if it is insensitive to situations in which one or more of its arguments are set to a value of zero or close to zero. Indeed, when the reliability of these values is equal to zero, their corresponding weights will be zero. Therefore, there will be no systematic error such as occurs in the typical case of averaging when the weights are constant.

Since the interaction between internal (model) and peripheral information flow is organized in the same way as between different components of peripheral inflow, the internal model of the controlled object can be regarded as an additional sensor with a particular magnitude of reliability. In the case of deafferentation, the sensory information processing system depends entirely upon the internal information source for determining the state of the controlled object.

A control system obviously has to possess a knowledge of the correlations between different afferent channels, including model channels. A neural network has such information in the form of conditional probability distributions. A sensory processing subsystem uses this knowledge to determine the most probable current state of the object by paying attention to the most reliable information channels. Optimum control influence sent to the object corresponds to the most probable current state of the object. A mathematical description of some of these abovementioned processes can be found in Section 1 of the Appendix.

4.2 The Essence of the Internal Model of the Controlled Object

In a general sense, a description of controlled object dynamics consists of a constant portion, which is mathematically represented in the form of a differential equation system, and a variable portion corresponding to the current values of phase space coordinates. It should be emphasized that the variable portion reflects a unique object property according to which the future state of the object depends not only on controlling actions, but also on its previous state.

A neural OCS receives information about the controlled object in the form of sensory afferent signals. What coordinate system should be used in the OCS for an internal representation of the controlled object state? This question is far from trivial. The situation resembles tensor calculus, in which properties of a material system are considered independently of the choice of a particular coordinate basis. In the nervous system, different coordinate systems correspond to different sensory systems (e.g, visual, auditory, tactile), which can provide information about one and the same object. What is the form of the information pattern that is common to all such systems and is understood as the object state? If we pursue the issue of OCS internal organization that is based on the complex structure of afferent projective fields, we take a step toward analyzing an OCS as a complex hierarchical system (see Chapter 7).

4.3 The Internal Representation of the Controlled Object Phase State

Up until now, we thought that any controlling neural system had an "intrinsic notion" of the phase state coordinates of its controlled object and used their values as midpoint results in the processing of sensory information. Control output and model "sensory" signals are calculated using such values. However, we should clarify whether the need for such a "notion" is real and objective, or whether it is merely a convenient way for us to describe a potential structure of a neural control system.

The necessity for internal representation of the knowledge of the controlled object state arises in the setting of incomplete object observability. The simplest example of such a situation is a necessity for calculating spike frequency during a nonzero time point in which the strength of a sensory signal is usually encoded. In the setting of incomplete observability, the value of a control action cannot be determined merely on the basis of the sensory information that is received at the current moment in time. The previous history of the controlled object's behavior should also be utilized for calculating control output. The use of short-term operative memory for just this purpose is, in fact, one of the potential forms of object state representation.

The function of the internal model of controlled object behavior is described in Section 2 of the Appendix. Section 3 of the Appendix describes how a probability-based approach may be used for the realization of function of "encoder-decoder" type in neural networks.

The ability to store previous history gives an OCS a rather interesting advantage. Obviously, the time necessary for transferring information regarding a new controlled object state should be as short as possible. On the other hand, the partial observability of the controlled object makes it necessary to store this information in the intrinsic memory of an OCS. If the OCS possesses a memory mechanism and hence can store the previous object state code, the code for a difference between stored and subsequent states can be transferred instead of the new state code. We will name the mechanism of encoding described above "differential encoding." The simplest example of such an encoding mechanism in the nervous system is the process of information transference between neurons by electrical impulses. The role of a memory mechanism is performed in this case by temporal integration of the excitatory postsynaptic potentials on the postsynaptic membrane. Longer memory can be produced by increasing the complexity of the postsynaptic neuron (i.e., mechanisms of long-term depolarization of the soma and/or the dendrite tree, and bistability of the cell state) (Arshavsky et al 1985d; Hounsgaard et al 1984; Hounsgaard and Kiehn 1985).

4.4 The Principal Features of the Neural Organization of Internal Representations of the Controlled Object State and Its Model

It is well known in pattern recognition theory that a recognition procedure consists primarily of so-called preprocessing and a procedure of decision making. During preprocessing, the original phase space of object criteria, which in general is not orthogonal, is transformed to the orthogonal space of secondary criteria, whose dimensions are usually much fewer (on account of the discarding of dependent criteria). Such transformation, as a rule, allows for significant simplification of the procedure of solution making, which is very important from the perspective of technical realization.

It was shown above that the necessity of forming an intermediate result of sensory information processing for an OCS is actual and objective. Is it necessary for the internal representation of the object state to be orthogonal, and if so, then in what sense? We may functionally represent the correlations between afferent signals as their corresponding conditional probabilities, which are stored in the neural network of the OCS in the form of synaptic weights. It takes a rather complex system to use these correlations and to perform operative synaptic weight setting. It is clear that if there were no such correlations between afferent signals, the neural network would be much less complex. So far as simplicity of organization is one of the OCS's criteria of optimality, the necessity for orthogonality of information channels in an OCS, in the sense of an absence of correlations between the corresponding signals, is actual and objective: i.e., independent of the material substrates of the OCS.

When full orthogonality is present, it corresponds to the maximum information density in an OCS network and, on the other hand, to minimum signal interference protection. One can speak about the optimum relationship between information density and the degree of interference protection. The probability of the occurrence of interference is obviously much higher in the periphery of the nervous system than in the spinal cord. Information density should therefore be increased significantly during the processing of peripheral afferent signals in the intraspinal OCS.

4.5 Neural Mechanisms for Calculating the Most Probable Current State of the Controlled Object

The most probable value of a random variable is equal to its mathematical expectation only if the corresponding probability distribution is unimodal. Otherwise the problem of choosing the largest of the distribution maxima arises. What is the anatomy of a neural network performing this function?

The intrinsic relationship between any two sensors or effectors that are determined by controlled object anatomy can be characterized as one of three types:

1. antagonistic, e.g., flexion and extension of the same joint
2. indifferent, e.g., sensors or effectors of different joints, limbs, etc.
3. agonistic, e.g., motor units or receptors of identical types belonging to the same muscle

Indifferent sensors or effectors may become functionally antagonistic or agonistic depending on the control task. Moreover, the relationship between them may change with the transition from one to another region of a movement trajectory.

Antagonistic effectors should not be activated simultaneously as a rule. This requirement arises from the fact that the control law must be optimum in the domains of energy and time expenditure. The solution to the optimal control problem in which all effectors are more or less activated, and each antagonist competes to move the controlled object in its own direction, results in a waste of energy. Therefore, the sensory information processing subsystem must provide maximum contrast between two controlled object state regions corresponding to activation of such antagonists by providing strong reciprocal inhibition between neurons activating these effectors.

The multiple inhibitory connections between neurons involved in control of activation of antagonistic muscles apparently reflect the negative correlation between their activities in a system with optimum performance. Synaptic weights should be changed when the network alters its working program, e.g., functional agonists must become antagonists or, alternatively, antagonists must be coactivated to reach optimum in certain cases. Such alterations in synaptic weights can be made, provided that the involved neurons can perform the multiplication of afferent signals, which in turn can be done with the help of complex synapses or specially structured dendritic trees.

While discussing the functional organization of the motor output of a neural network, we noted the necessity (originating from the optimality request) for a stochastic realization of the optimum control law. The process of recruitment/derecruitment of motoneurons in the spinal neural OCS is a good example of such organization. A neural mechanism underlying this process must provide stability of definite integral value of the control action. It is possible to show that such a mechanism can be based on reciprocal inhibition between motoneurons belonging to the same muscle (e.g., through Renshaw cells).

One might consider the scheme of reciprocal inhibitory interaction to be a simple and widespread example of the existence of internal feedback in a neural OCS. Presynaptic inhibition through the PAD generating system may also be regarded as a variant of reciprocal inhibitory interaction between sensory information channels. Practically speaking, a major portion of central pattern generator descriptions and models is based on the scheme of reciprocal inhibition. It should also be noted that all these schemes are aimed at imitating the ability of a real system to generate cyclic activity patterns. The authors describing these schemes do not, unfortunately, explain why the patterns that are generated should be as they are. They regard the neural network of a generator as a program control mechanism and not as a control system with feedback. There is no room in such models for peripheral feedback and, consequently, the issue of uniting the program with feedback types of control is not approached.

From the above discussion, a question emerges regarding the possibility of pointing out exactly which neurons in the spinal cord store information about the current state of the object controlled by the spinal OCS. Attempts to establish such simple relationships between function and structure should be approached with great caution. A motoneuron, for example, which seems at first glance to be a "pure" output unit, may at the same time be regarded as a sensory neuron because it receives direct input from muscle Ia afferents. Therefore, the shortest internal feedback circuit from a motoneuron to itself through inhibitory interneurons (Renshaw cells) should be considered as a channel for transferring model information flow from the control output of the spinal OCS to its sensory input.

Thus, motoneurons should be included in any neuronal group encoding the control object state. There are large numbers of interneurons that receive inputs directly from peripheral afferents or from other neurons, including motoneurons. Information processing in the brain may be

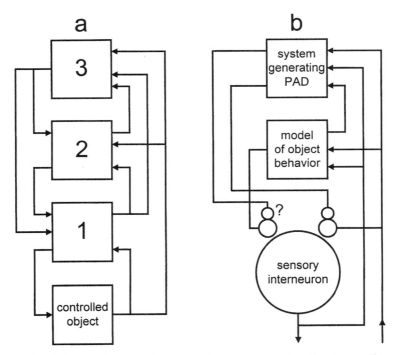

Figure 39—Schema of neuronal organization of the spinal optimal control system (OCS). a, interaction between OCS subsystems and the controlled object; 1, 2, 3, different interneuronal levels; b, interaction between peripheral and model flows.

conceptualized as a translational process that proceeds from one descriptive language of the controlled object to another. From this point of view, sensory information processing in the spinal OCS may be represented schematically as a translation of information into the "native" language of Ia afferents, which "talk" directly to output motoneurons.

It is necessary to underline once more that one cannot precisely identify the location in the spinal OCS neural network where the information about the controlled object (limbs as a mechanical system) is transformed into controlling signals. The first stage of that transformation is dictated by the structure distribution of sensors on the controlled object, for example, the anatomy of connections between muscles and their skeletal framework for muscle afferents. The final stage, the application of muscle forces to the limb skeleton, is obviously determined by the same anatomy described above.

A very simplified scheme of the neural structure of the spinal OCS is shown in Figure 39. Many important organizational issues for this system (e.g. motor program initiation, the structure of stretch reflex system) are not reflected in this figure. These issues will be considered below.

5

Generalizing the Concept of a Neural Optimal Control System: A Generic Neural Optimal Control System

The representation of the spinal motor control system in the form of a neural OCS can explain all major experimentally revealed peculiarities of the structure and function of this system. Now it is necessary to generalize the idea of a neural OCS in such a way that it can be applied to any subsystem of the nervous system.

The reader has probably noticed that we have not explored a very important issue regarding mechanisms of motor program initiation. According to the concept of an optimal control system described above, motor program initiation may be regarded as aim setting. In the simple case of the scratching program, the control task of limb movement is described by the target coordinates of the aim state to which it is necessary to bring the controlled object. This concept provides a rather simple yet comprehensive explanation for the fact that motor program initiation can be performed by comparatively simple organized tonic signal information flow descending to the spinal cord from higher brain structures. However, in the case of locomotion, it is not clear enough what to consider as the aim state, since the controlled object is first a limb, in the swing phase, and then the limb together with the body, in the stance phase. The complexity of the initiation issue reflects the fact that it concerns one of the least investigated domains of movement physiology.

Another aspect of the concept of an OCS is the basis for determining optimality criteria. The tendency for minimization of energy and time consumption is natural for controlling a massive inert object. But this criterion seems hardly appropriate for all brain subsystems.

Finally, the third aspect requiring generalization is the idea of a controlled object. We have previously considered a limb of an animal as such an object. Signals descending from suprasegmental structures are regarded as initiat-

Biological Neural Networks
Konstantin V. Baev
© 1998 Birkhäuser Boston

ing, or carrying information about, a control aim. Descending afferent flow also supplies the spinal OCS with information about the limb position in relation to gravity, obstacles, and so on. Moreover, stimulation of various suprasegmental structures during fictitious locomotion or scratching evokes changes in generator efferent activity, which are—as one might anticipate—very similar to those observed in the case of stimulation of limb afferents (Degtyarenko et al 1990, 1992; Degtyarenko and Zavadskaya 1991). Thus, we should in general understand "controlled object" as the whole system of afferent information sources and controlling signal receivers.

The analysis of a possible means of generalizing the OCS concept reveals inextricable relationships among all three of the issues discussed above. If information about the control aim is considered to be emanating from the controlled object, how does it differ from any other information that describes the current object state? From a mathematical perspective, it is convenient to choose an equation system concerning object state variables as a general form of the control task description. The goal of any control task consists therefore of bringing the controlled object to a state that satisfies the equation system, although the set of solutions for this system can be rather complex. According to such a description, we can designate initiating signals as information about the current values to which the left part of the equation equals. Thus, *initiating signals differ from other afferent signals only in that the OCS tends to minimize certain integral measures of these initiating signals.*

This representation of a control task is very useful, as it allows one to generalize simultaneously the ideas of program initiation and optimality criteria. Is the criterion of minimization of energy and time consumption a unique case of a more general criterion? The time consumption criterion can be obtained from the general criterion when the initiating signal takes on only one of two values: nonzero and zero. The issue of the energy consumption criterion is less trivial. To adequately address it, we should define what an initiating signal source can be.

We have defined initiating signals as sensory signals of a unique modality. Therefore, it is natural to propose the existence of special sensors that are the sources of initiating signals. Obviously, such sensors cannot always be a receptor (e.g., there is no "receptor of enemy"). As animals accumulate experience, very complex neural networks can be formed that are able to recognize, for example, images of prey, enemies, or possible sexual partners, whose appearance serves as a signal for initiating an entire behavioral program. However, not all sources of initiating signals are formed exclusively during new experiences: an initial number of such sensors must be set according to an inborn genetic program. Otherwise, the animal would simply perish because it has not had enough time to learn. Moreover, it would not be able to learn without any initiating signals. A typical example of a genetically-predetermined system of initiating signal sources is the pain sensitivity system.

We do not know whether there are special receptors that provide information about the rate of energy consumption, energy deficiency, and so on. If not, such information can apparently be derived in the course of processing sensory inflow. We will consider in Chapter 7 how the sensors-detectors of complex patterns might be constructed during the accumulation of experience in the form of neural networks on the basis of more simple and primary sensors.

Thus, according to our generalized concept, any set of afferent signals may be divided into a subset of initiating signals and a subset of noninitiating signals that we will refer to as "signals of information context" (or simply "informing signals").

Therefore, now we can make the following generalization. Every OCS is constructed according to the same functional principle, regardless of its location in the hierarchy of the nervous system (Figure 40). Such neural controlling systems contain two distinct functional subdivisions: (1) a controller, which is the subsystem providing a governing set of rules or commands—a *controlling law*—that directs the action of the recipient of these rules—the *controlled object*, and (2) a model, which is the subsystem that generates a *model* of object behavior, i.e., the expected afferent flow from the controlled object. The function of the model is to predict with high probability any given state of the controlled object that can result from influences on it from its controlling system.

The necessity of the presence of a model was already discussed in Chapters 2 and 4. However, it would be useful to briefly mention this again. There are two major reasons for the controlling system to have a model of object behavior. The first is the feature of incomplete observability of the object. This feature is best demonstrated for higher levels of the nervous system, although it is true for any level. For instance, the visual system receives information from only a portion of the environment. Incomplete observability ultimately means that channels of afferent information are noisy ones. The second is the feature of incomplete controllability. In most cases the controlling system cannot bring the controlled object into a new state in a single control step, and the results of any intermediate computations must be stored in a controlling system, in order for it to perform recursive computations. In addition, it is also necessary to mention the existence of significantly elongated temporal intervals between the departure of a controlling signal and the time of arrival of the corresponding afferent information from its controlled object in the control system. For instance, in the case of cortically-initiated movements it could be tens of milliseconds.

The presence of a model within the controlling system gives it certain advantages. Model afferent flow is used by the system to determine the most probable current state of the object by using a filtering mechanism for noisy

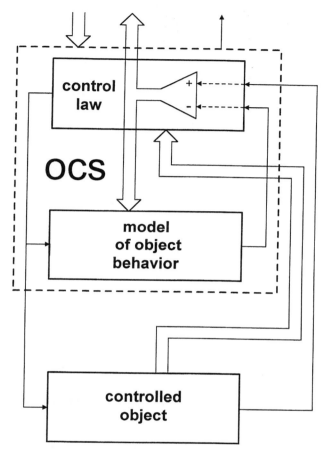

Figure 40—Functional architecture of a generic neural optimal control system (OCS) and the types of signals that the controlled object and subunits of the OCS send to each other. Informational and initiating signals are shown by single-line and double-line arrows respectively. See text for explanations.

afferent information. The complexity of the filtering processes can vary significantly in different neural systems, but because of the probabilistic nature of neural networks, the basis for this process is usually a knowledge of the conditional probability of an event. This knowledge is stored in the network in the form of neural connections and individual neuronal properties. Model and real flows interact on a parity basis. For simplicity, we shall place an optimal filter in the control law box (Figure 40).

Determining the most probable current state of the controlled object was mentioned above because it is the most frequently encountered situation.

However, it is necessary to mention some other possible objectives, such as determining the mathematical expectation, median value, etc., of the object being in a particular state. Complex control systems are capable of utilizing different criteria to determine their object states.

Another advantage of having a model is the ability of a control system to receive crucial information such as mismatch or error signals. A mismatch signal between real and model flows is necessary for the process of learning (see below) and is also computed during the filtering process (Figure 40).

Two functional subdivisions of the generic OCS were described above. Anatomically they may be inseparable. Both the control law and the model can be realized within the same neural circuit in the simplest of control systems. For example, some CPGs can consist of only one neuron that performs both controlling and modeling functions. In more complex cases, such as the cortico-basal ganglia-thalamocortical loop (see Chapter 11), these functional subdivisions can be more clearly separated anatomically because a more complex neural network is necessary to predict the behavior of such a complex object as the body and the environment.

Any functional neural system involved in the expression of a discrete automatism has an output through which it sends controlling signals to a controlled object (effector organ), and two types of afferent signal inputs that contain the information necessary to compute an output signal: (1) *initiating* signals and (2) signals containing the current *informational context* in which a system as a whole finds itself. The system attempts to minimize initiating signals—or, strictly speaking, to minimize the integral measure of initiating signals—by using informational signals to compute proper output. Initiating signals can be considered analogous to "energetic" signals. They activate systems, resulting in a realization of any system's corresponding automatism. Given the concepts mentioned above, one may now formulate the following definition of an automatism: An *automatism* is a program of action or sequential actions that is stored in a neural network and that, during execution, leads to a minimization of its corresponding initiating signal. Withdrawal or escape reactions and locomotion are typical examples of automatisms. They serve to avoid danger, i.e., to minimize a signal about danger. However, one may easily cite numerous examples in which the final goal of an automatism is *maximization*, not *minimization*, of initiating signals, such as during pleasurable reactions when the corresponding signals are maximized. There is no contradiction, because the nervous system can easily invert the sign of the signal so that it becomes minimal instead of maximal.

It is necessary to clarify some other aspects of the classification of initiating and informational afferent signals. Appropriate segregation of afferent flow into these two types of signals can only be made at the level of the recipient OCS. The same control signal of one OCS can be interpreted differently by other OCSs. They may be either initiating or informational signals. This concept will be explored later with different examples of specific neural networks.

Initiating and, of course, informational signals can be short- or long-lived, depending on the particular control task. The latter is the case when the controllability of the object is very low, and it takes a longer time to remove a corresponding initiating signal by sending complex control influences to the controlled object. Finally, there are subtypes of initiating signals. For instance, there is one that starts an automatism, and there is another (a mismatch between model and real flows), an error signal, which goes to the model subsystem of a network and to the higher level (see Figure 40). Both have to be minimized during control. Therefore, a mismatch signal is the initiating signal for learning processes within the model, i.e., it initiates a learning automatism. Below, we will try to describe each signal as precisely as possible, to avoid confusion, especially in those cases when a mismatch signal is sent to another OCS.

Let us consider some mechanisms of mismatch computation in the nervous system. It is necessary to start by analyzing the functional abilities of a single neuron in terms of its state invariability in relationship to input signals. The simplest mechanism providing such a property is, apparently, the mechanism of threshold adjusting to the intensity of input signals. By means of this mechanism, the excitation threshold somehow "models" the average level of membrane depolarization. In other words, a neuron possessing such a mechanism calculates a mismatch between the total afferent inflow and the "internal model" flow, a role played by a threshold value. In a single neuron, an internal model of the environment can also be used functionally, as described in Chapter 4, as an additional source of signals taking part in optimal filtering of afferent information, which significantly increases the reliability of the system. This type of functional organization is peculiar to pacemaker neurons.

A mismatch signal, according to its functional meaning, can in general be negative or positive. Since the quality of a signal in the nervous system is encoded mostly by the site of its source, a system of at least two neurons is necessary to produce a mismatch signal in its general form. A signal from one of those neurons can be regarded as a positive mismatch, while that from another can be regarded as a negative mismatch. It is clear that the neurons should not be activated simultaneously, a situation that can be easily avoided by reciprocal inhibition. A well-known example of such a double-unit system is a trigger, the simplest memory unit having two stable states. A double-neuron system can apparently work as a trigger. However, a third state of the system is possible, one in which neither neuron is activated. This situation corresponds to the absence of mismatch (information about mismatch is unreliable). Furthermore, all of the states should not necessarily be stable.

Two major neural schemata used in the brain for calculating mismatch are shown in Figure 41a, b. These can be united into a more complex scheme, since they can be combined into a complex neural network if necessary. Let us now consider their functional properties.

The schema shown in Figure 41a produces no output signal if the input signals are equal to each other and alter simultaneously. A typical example of such a real schema is present in the motor output level of the spinal cord, where a-motoneurons and inhibitory Ia interneurons are included in the circuitry.

In the schema shown in Figure 41b, differentiation of simultaneous changes in input signals is performed. In addition, the existence of feedback allows (under corresponding values of synaptic weights) the schema to work as a trigger, i.e., a memory unit. Thus, such schemata can be used for information transfer with differential encoding. A typical example of such a schema is the backward inhibition of granule cells through Golgi cells in the cerebellar cortex.

An analysis of the structure of brain neural networks leads one to the conclusion that mismatch signals are widespread in the nervous system, since the two neural schemata considered above are actually the major structural units found in the brain. Mismatch signals entering one or another neural system may be initiating or information signals from the "viewpoint" of a given system. Such signals can, apparently, duplicate each other and should obviously be processed as usual. They can undergo optimal filtering, or a signal of mismatch between them can be calculated. The schema of parity interaction between two mismatch signals may be the same as described in Chapter 4. The mechanism of operative attenuation of synaptic

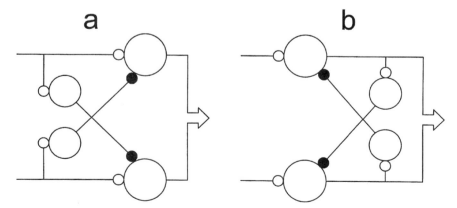

Figure 41—Examples of neuronal schemata for calculating mismatch signals. Big circles, neurons; small white and black circles, excitatory and inhibitory synapses, respectively. Explanations are given in the text.

weights is functionally necessary in both cases. Thus, such a schema may be considered as a substrate of functional organization for the brain and has a higher level of complexity than a single neuron.

The term *controlled object* also needs some additional explanation. It is possible to talk about a controlled object by using either a limited, narrow definition or a more broad and general description. In the first case, it refers to the executive organ itself or to a lower OCS when one is describing a controlling system that is present at a higher level in the hierarchy of the nervous system. In the second case, it refers to the objects defined in the first description as well as to all surrounding interconnecting neural systems to which a given OCS sends efferent signals and from which it receives afferent signals. Ultimately, it refers to the whole surrounding brain, when the highest functional levels of brain activity are under consideration. At first this sounds strange because any OCS attempts to create a model of object behavior. For a lower OCS, it is the task of trying to predict the behavior of the whole brain that seems impossible. Nonetheless this is true; however, because of limited computational abilities, such an OCS is capable only of restricted predictions about the higher systems. For instance, a spinal OCS is not capable of predicting descending influences generated by higher brain levels on the basis of visual information. This broad understanding of a controlled object has profound consequences: (1) there is no way for a lower OCS to determine the existence of higher functional levels, and (2) a higher-level OCS has to "speak" with a lower one by using the language of the lower level, i.e., the language of lower-level initiating and informational afferent signals. In the discourse to follow, the more limited and narrow understanding of the controlled object will be utilized.

Obviously, while performing an analysis of the function of any part of the brain, one can separate the whole controlling system into OCSs and their controlled objects. On the other hand, one can combine numerous OCSs into one complex system and also consider it as an OCS. In both situations, the method will succeed if an OCS and its corresponding controlled object are designated correctly, i.e., if a *functional* approach is used.

6

Learning in Artificial and Biological Neural Networks

6.1 The Problem of Learning in Neurocomputing

The dominant biological approach to the problem of learning was discussed in Section 1.2. We saw that, in neurobiology, emphasis is put on the experimental proof of long-term changes in neural network parameters, mainly parameters of synaptic transmission. In neurocomputing, the approach is completely different.

The main idea currently developing in neurocomputing is simple and quite captivating. It is the idea of training a system instead of programming it to carry out information processing. This idea has influenced the approach to the problem of learning in neurocomputing. In this approach, changes in synaptic weights are considered to be natural and necessary conditions for neural network learning, and the focus of attention of scientists, mathematicians, physicists, and technologists is directed toward finding the most effective algorithms of training. It is necessary to note that the term "training," itself, implies the presence of a teacher in the process of learning.

Kolmogorov's mapping neural network existence theorem and backpropagation neural networks are the foundation of neurocomputing. The architecture of a backpropagation neural network consists of K ($K{\geq}3$) layers of processing elements, and in some respects is quite similar to Kolmogorov's network. Layers from K$=$2 through $K-1$ are also called hidden layers. The first layer consists of n fanout processing units that accept components x_i of the input vector and distribute them without modification to all elements of the second layer. Each unit of each subsequent layer receives output signals from each element of the layer below. This process continues until signals reach the final output layer. The output K^{th} layer of elements consists of m units and produces the network output vector. This output vector is considered as a networks estimate y' of the correct

Biological Neural Networks
Konstantin V. Baev
© 1998 Birkhäuser Boston

output vector **y**. In addition to feedforward connections, each unit of the hidden layer receives "error feedback" connections from each element of the layer above it.

The backpropagation network operation cycle consists of two sweeps. The first sweep starts by inserting the **x** input vector, and finishes by emitting an output estimate **y'** (the forward pass). The second sweep (the backward pass) starts by inserting the correct output vector **y** into the output units. During this backward pass, the modification of processing element weights ("synaptic" weights) takes place. During the training of a neural network, it is much more convenient to change "synaptic" weights than to change other parameters of network elements. Thus in the next cycle, elements—more exactly, some of these elements—will have new transfer functions, and a new adjustment of network parameters may be performed. During such multistep iterative training processes, it is possible to implement a specific function to any degree of accuracy. Despite the fact that the corresponding theorem has proven that a hypothetical three-layer backpropagation neural network is adequate for the approximation of real-world functions, in practice it is better to have four, five, or even more layers.

Different backpropagation learning laws were subsequently proposed. There are also many other learning laws that can be used to train neural networks. The training regimens used in neurocomputing can be divided into three major types: (1) supervised training, (2) graded or reinforcement training, and (3) self-organization.

Supervised training implies a regimen in which the neural network receives a sequence of pairs of input and correct output vectors. Frequently, these pairs are assumed to be examples of a fixed function f, but the relationship between input and output vectors may also be stochastic.

Graded training differs from supervised training only in that, instead of being given the correct output vector for each training trial, the network receives a grade or score. This grade tells the network how successful it has been over a sequence of training trials. Graded training may also be periodic or aperiodic.

During self-organization, a network modifies itself only in response to **x** input. R. Hecht-Nielsen (1990) writes that "this category of training may seem rather pointless, but a surprising number of information processing capabilities can be obtained using it." The reader who wishes to obtain additional information about the current state of neurocomputing as a discipline may find it in the brilliant recently published textbook of R. Hecht-Nielsen (1990).

It is worth mentioning that in neurocomputing, training usually starts from scratch. In other words, neuron-like elements are usually symmetrically connected at the beginning of a learning session, and learning should occur in a reasonable number of trials.

6.2 Basic Principles of Learning in Biological Neural Networks

An ideal way to approach the problem of learning also utilizes the concept of an *automatism*. If one attempts to address the question, "What is learning in its functional sense?" using the concept of an automatism, the answer is rather simple. Learning is a process that leads to the appearance of a new *automatism*, or to the improvement or refinement of one that already exists. In this case, memory is necessary to fix within the network the new computational decision and its associated automatism. But immediately a new question appears: According to what principle is learning built into neural control systems which, as mentioned previously, are optimal and contain a model of their controlled object?

It is intuitively understandable that common principles of learning in neural control systems should exist, but the mechanisms associated with their realization may differ significantly in different animal species, and even in different parts of the nervous system of the same species. It is also understandable that plasticity and learning are principally similar processes, but they have different temporal durations. Moreover, it is necessary to consider the evolution of the nervous system (and evolution in general) as the evolution of *automatisms*, i.e., as a *learning process*. With such an approach, the realization of any neuronal automatism may be conceptualized as the reading of information from the network memory that was established within the neuronal network in the course of evolution and in the life-long learning process of the individual.

In neural control systems, learning occurs according to the following schema. The network subsystem realizing control laws and the network subsystem realizing the model of the controlled object adjust their parameters, and hence their computational functions, in such a way that mismatch signals are minimized. Therefore, learning starts when a mismatch signal increases, and finishes when it is minimized.

If one analyzes the learning laws used in neurocomputing and computational neuroscience, it will also be apparent that this principle works in cases of supervised and graded training. Specific criteria are always used, for which minimization is the desired result of the learning procedure. However, there is a radical difference between learning in artificial networks and in actual biological neural networks. As mentioned above, learning usually starts from "scratch" in the first case. This is why major efforts have been made to find the most effective learning laws, without which the network quite often reaches a local minimum on the *error surface* and stays there despite continued learning. Biological neural networks are genetically predetermined to work with a specific class of functions, and subsequent learning proceeds rapidly in the real situation.

From a mathematical point of view, learning should be considered as optimization. Criteria of differing complexity can be minimized during this procedure. Situations in which it is necessary to minimize a specific parameter are well known in physics, such as the *principle of variation* and the *principle of least action*. As mentioned in Section 2.2, in control theory some control parameters are often optimized as well, and if numerous criteria are minimized, it is customary to talk about a *global minimum* that is reached by the system.

In most cases, learning in nature occurs without a teacher. Therefore, the main question is: how can a network find a correct decision without knowledge of the correct answer? There must be a means by which a network can do this. Otherwise, natural neural networks would not exist. The basic mechanistic principle by which this occurs has to be universal and permit further improvements.

As a first approach to this problem, we will focus on this basic principle of the learning process, regardless of its rate of development. It does not matter how long it took to develop such learning in the course of evolution. Nature had millennia and a huge laboratory in which to develop this capacity.

Given the idea that each neural control system receives two types of afferent inputs—initiating and informational—it is quite easy to formulate the basic principle of learning (Baev and Shimansky 1992; Baev 1994, 1997). It consists of random searching in hyperspace, every point of which is a control system with a unique structure and set of values for its parameters. The intensity of the random search is proportional to the intensity of initiating signals. In other words, the network will try to modify its computational function in such a way that an initiating signal will be minimized by the end of the learning process. In order to imagine this type of learning, let us consider a few simple examples.

6.2.1 An Analogy: Brownian Motion of Particles in the Presence of a Temperature Gradient

The concept of Brownian motion is well known to all students of science. It is the result of random collisions of a particle with surrounding molecules. The movement of a particle toward any direction in space has equal probability, and the average amplitude of movement is determined by the kinetic energy of the surrounding molecules, i.e. by temperature. In the absence of a temperature gradient, a particle will search infinitely in a volume of liquid or gas, and the probability of its presence at any given point within that volume will be the same. But in the presence of a temperature gradient, sooner or later the particle will reach a point having minimum temperature and will not be able to leave it because the kinetic energy transmitted to it by the surrounding molecules will be minimal at this point, and the particle will not leave its location. The results of the modeling of this simplest case are shown in Figure 42a. The essence of this simple model consists of the following: at each step of calculation, the shifting in either a positive or a negative direction along the x and y axes is equiprobable. The amplitude of shifting is set by a generator of random numbers and is multiplied by a value proportional to the distance between the current position of a particle and its destination point. Therefore, the average value of this amplitude decreases as the particle approaches its destination point. This simple example demonstrates well the regular principles of the process: (1) The steeper the dependence of the amplitude of shifting on the distance between the current position of the particle and a destination point, the faster the particle will reach the vicinity of the destination point, and (2) the higher the frequency of a random search, the faster the particle will reach its destination point (this is not shown in Figure 42a). Hence the state of such a system will be stable when the particle is in the vicinity of its destination point.

Suppose the system described above could determine the temperature gradient, and then move the particle toward its destination point along this gradient. The result is shown in Figure 42b. It is quite obvious that the particle reaches the destination point much faster.

6.2.2 Change of Neuronal Transfer Function Due to the Influence of Initiating Signals

Let us consider a hypothetical neuron receiving informational and initiating inputs. Suppose there is an intracellular mechanism increasing the fluctuations of the synaptic weight of the informational input when there is an increase in the intensity of the initiating signal. Both directions (increase and decrease) of change in synaptic weight are possible. Suppose we want to change the initial transfer function y_i to y_d (desired destination) by using the above-described mechanism. It is then necessary to create an initiating signal that will be proportional to the difference between the current transfer function y_c and y_d and to send this signal to the neuron at each step of the calculation. If these two functions differ only in synaptic weights, then it is enough to adjust only the synaptic weight to receive a new transfer function. The corresponding synaptic weights are w_i and w_d. It is clear that if the intensity of the initiating signal is proportional to the difference between current and destination synaptic weights, then, with time, a new synaptic weight w_d will be established, and this new state of the neuron will be stable because the initiating signal will now have a minimum amplitude. The results of modeling such a unidimensional search are presented in Figure 43. Thus, this system is very similar to Brownian movement in the presence of a temperature gradient. The initiating signal may be considered to be an "energetic" signal in this case. Obviously, the process can go much faster if the system can determine the gradient. In this case, with the same

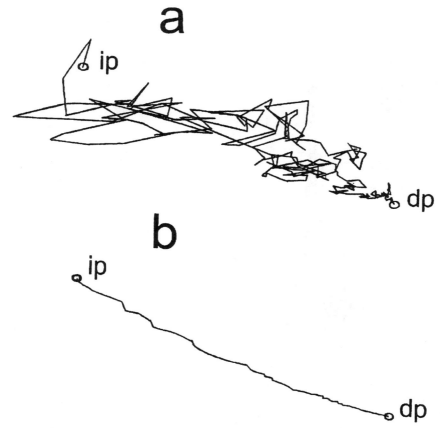

Figure 42—Random search of a Brownian particle in the presence of a temperature gradient (a), and the behavior of a similar system capable of determining the direction of temperature gradient (b). Temperature at the destination point (dp) is lower than at the initial point (ip).

time scale as in Figure 43, the curve of synaptic weight adjustment coincides with the y axis (not shown in the figure). A specific temporal relationship between the initiating and informational signals should be present to make this scheme of learning work in a real neuron. These relationships will be discussed below.

6.2.3 Change of a Function Calculated by Neural Network

It is easy to extrapolate the above-described process to a neural network receiving two types of afferent inputs—informational and initiating—and

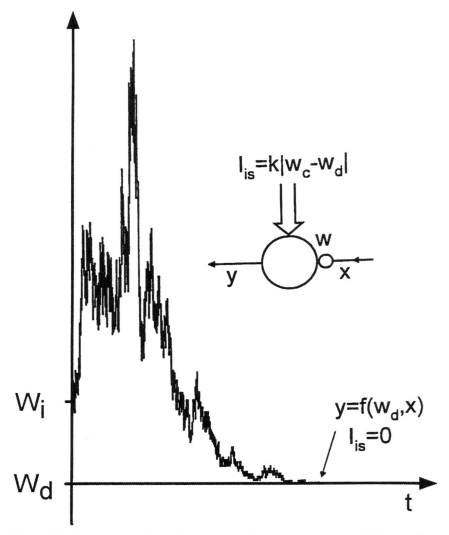

Figure 43—The change of transfer function of a neuron under the influence of an initiating signal. w_i, w_c, and w_d, initial, current, and destination synaptic weights, respectively; x, y, input and output signals, respectively. Intensity I_{is} of the initiating signal is proportional to the difference between current and destination synaptic weights. t is time. Adjustment of synaptic weight stops when I_{is} equals zero.

realizing a specific controlling function. Such a system is also able to perform random searches in the hyperspace of its states under the influence of an initiating signal. The dimensionality of hyperspace is equivalent to the number of parameters that may be changed under the influence of an initiating signal. Below we will consider only the example of the case of synaptic weight changes in the course of learning, although it is clear that other crucial parameters of the system may change under the influence of initiating signals.

If such a neuronal system has several layers (three or more) and feedback loops, then, as was shown in Section 2.3, it has the potential ability to calculate rather complex functions. If the function responsible for minimizing an initiating signal is included in the class of functions that may be calculated by this network after adjustment of synaptic weights, then it is obvious that random searching will stop when the network begins to calculate this particular function (Figure 44). Although such rudimentary memorizing mechanisms as trace processes, the declining phase of postsynaptic potentials, afterhyperpolarization, etc., can still be present, in this section we will consider a network in which the special mechanisms of memorizing

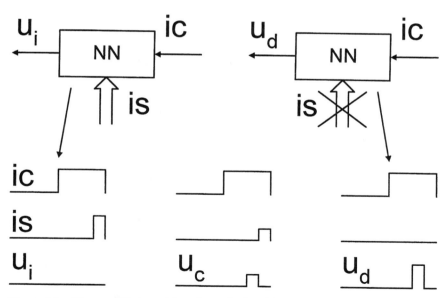

Figure 44—Changes of control function calculated by the neural network under the influence of an initiating signal. NN, neural network; u_i, u_c, u_d, initial, current, and destination control outputs, respectively; is, initiating signal; ic, informational context. Random searching initiated by the initiating signal stops when the signal becomes equal to zero i.e., when the control output is generated prior to the initiating signal. See text for explanations.

initiating and informational signals are absent (see Chapter 12). In this case, several specific conditions should be satisfied:

1. Informational signals should correlate with initiating signals and should appear in the network system earlier in time.
2. The influence of an initiating signal should be directed to those neurons and synapses whose activity was evoked by the preceding informational signals. In other words, a discretely directed backward time window extending from the moment that the network received its initiating signal to the moment the network received its informational signal must exist. Synapses active in this time window are the sole targets for initiating signals. Despite the presence of initiating signals of identical intensity, bigger fluctuations will be observed in more active synapses. It is obvious that there is no sense in changing the synaptic weights of less active or inactive neuronal connections.
3. The more frequent the coupling of informational and initiating signals, the more intense are the fluctuations of synaptic weights. In other words, influences of initiating signals are accumulated in the system.

Obviously, specific molecular mechanisms have to be responsible for such phenomena. The essence of these conditions will become clear later when specific illustrative situations are analyzed.

Two very important assumptions were made earlier. The first was that a network is not capable of memorizing a long previous history. The informational context necessary for the computation of a controlling output has to be present during the arrival of an initiating signal, or at least trace processes evoked by informational signals have to occur during the arrival of an initiating signal. Otherwise, this type of learning will not work. The second was that the so-called credit assignment problem, a well-known limitation of reinforcement learning, has to be solved in a network undergoing such learning processes (Minsky 1963; Barto et al 1983). This problem is how to deliver an error signal to the appropriate neurons and, consequently, to their corresponding synapses (spatial credit assignment) during a uniquely appropriate window in time (temporal credit assignment).

The method of random searching is universal because it gives a neural network the ability to reach a new decision when there are no preexisting algorithms to appropriately direct its output. But being universal, it may be very slow when a search is performed in the hyperspace of a network's parameters. Methods found by nature to accelerate this process by solving the credit assignment problem, and by making other improvements in function, were discussed in a previous paper (Baev and Shimansky 1992). For the purposes of this book, it is enough to briefly mention them. These include: (1) determination of the gradient of an initiating signal (see Figure 42), (2) adjustment of system parameters in a dependent fashion (pattern adjustment), and (3) memorizing incorrect decisions (possible at highest levels—see Chapters 11–13). In evolution, systems based on random search methods could find a rule or a set of deterministic rules to adjust their computed functions for certain situations. For instance, an initiating signal may possess a sign—may be excitatory or inhibitory—and show the direction of any necessary synaptic changes. It is obvious that any improvement in learning should be based on the specifics of a network as well as on its cellular and molecular mechanisms. Therefore, the more complex the strategy involved in the search for new decisions, the more complex should be the system's mechanisms.

The existence of functional modules that describe the behavior of specific functionally isolated parts of the controlled object, and are responsible for control of one object parameter, may also be a factor that significantly accelerates the process of learning (Baev 1994, 1997; Baev and Shimansky 1992). The elements of such a module should receive mismatch signals that can be minimized by changing the appropriate function calculated by the module. The informational context of the module should include all functional modalities that are necessary to solve its various control tasks. Moreover, the module can be created in such a way that it receives information about the direction of synaptic weight changes (see next section).

It is clear that situations during which different initiating signals with the same informational context are minimized demand a calculation of different functions. Different functions should also be calculated by a network receiving the same initiating signals and different informational contexts. When a network completes its computation, it reaches a unique and specific stable state in its state space. This final state is usually a state to which all nearby states will eventually evolve. These stable states are often called point attractors. Obviously, if a network learns how to approximate several functions, it means that a corresponding number of point attractors will be created. This is why a network with a substantial number of point attractors is usually regarded as having an associative memory. Retrieval of information about the location of point attractors from such a system can be accomplished from partial information—for instance, by sending a descending initiating signal that was present during a previous learning session and determined the task for the session for which current memory retrieval is now desired. Neural control systems that are *recurrent networks* and possess a model of their controlled object can also produce necessary control output without peripheral feedback.

6.2.4 Basic Principles of Classical Conditioning

The concept of establishing new neuronal routes during learning was discussed in Section 1.2. Other conceptual mechanisms can be found in the neurobiological literature. For example, different principles were developed by Gallistel (1990) in his recently published monograph. In it he analyzes the formal structure of the environmental relationships that animals master to time and orient their behavior, and he shows which representations different animals can and cannot compute, and the very nature of the computations that are used by such animals to derive these representations. Using these facts, Gallistel compellingly proves that, rather than limit themselves to the concept of formation of associations which has been the basis of traditional learning theory, scientists should begin to consider numerous additional learning phenomena within a framework that utilizes the concept of computation and storage of quantities. However, Gallistel does not demonstrate by what principle these computations occur in neural networks.

As mentioned in Section 1.2, it is customary to distinguish two forms of conditioning—*classical* and *instrumental* conditioning. In this section, only the first type of conditioning will be discussed. The second form will be analyzed in Chapter 12. Given the conceptual description presented above, it is possible to provide a natural explanation for classical conditioning if conditioned and unconditioned stimuli are considered to be informational context and initiating signal, respectively.

Because of its broad use in the neurobiology of learning, the term *association* needs some explanation. In the case of classical conditioning, it means that the response that was evoked by an unconditioned stimulus in the beginning of a learning session is evoked by a conditioned stimulus by the end of a learning session. It usually implies that the unconditioned component of the reflex becomes a part of a more complex conditioned reflex after learning has occurred, i.e., it becomes *associated* with the conditioned stimulus. The term *associative* has a similar but simpler meaning in neurocomputing, and designates a network that has learned to produce a correct response at the end of a training session that consisted of presenting pairs of associated input and output vectors. As we shall see, there is a serious difference between these two meanings. In most cases, *no* association can occur in a neurobiological sense during classical conditioning; the desired output is computed *de novo*. However, we shall use the terms association and classical conditioning according to their common usage in neurocomputing when an input becomes *associated* with a correct output.

Let us imagine how learning, as described in the previous section, can work in a real situation. Suppose a neural network repeatedly receives from its controlled object initiating and informational signals. They are correlated in time, and the latter signal precedes the former one. The initiating signal informs the control system that the object is in a dangerous zone of its state space, such as when the object receives a painful stimulus.

Any information preceding the initiating signal, such as information from tactile receptors or such distant receptors as antennae, for example, may function as the informational context associated with learning. It is obvious that such a primitive system will be in the state of random search, i.e., it will learn to calculate a new function, until the controlling output that optimally minimizes the initiating signal appears in response to the available informational context (Figure 45). This controlling signal (for instance, a motor command to a muscle) will first move the object from the danger zone, and an initiating signal will either not be generated or will be minimal. It is understandable that such a system will compute different functions depending on the informational signal that preceded the initiating signal, the features of the controlled object, and the parameters of the initiating signal.

The learning strategy used for classical conditioning is shown in Figure 46. An initiating signal itself can produce an unconditioned response at the start of a learning session. By the end of the learning session, a con-

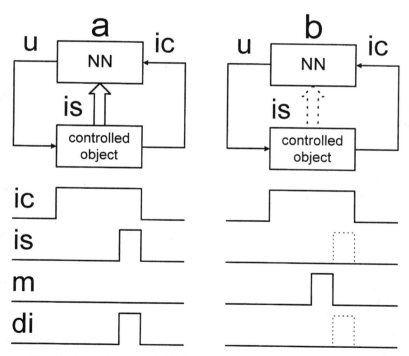

Figure 45—Minimization of an initiating signal received from the controlled object. a, b, signal exchange before and after learning, respectively; m, movement; di, dangerous influence. Other designations as in Fig. 44.

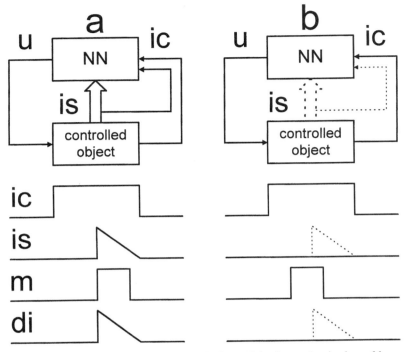

Figure 46—The simplest scheme of classical conditioning. a, beginning of learning session; b, end of learning session. Designations as in Figs. 45 and 44. See text for explanations.

ditioned response occurs before the arrival of the unconditioned signal that functions as an initiating signal. Obviously, neither of these learning mechanisms is *associative* in a biological sense. At the onset of the learning procedure, the system does not know what to compute, and there is nothing with which to associate an input. The timing of the process reveals that the conditioned response is calculated *de novo*. It does not appear as though the conditioned stimulus evokes the unconditioned response after learning has occurred. However, higher brain levels can evoke such responses (see Chapter 12).

At the same time, while computing an essential output *de novo*, a controlling system can take advantage of the knowledge of the direction in which synaptic weights must be changed. Let us consider several theoretical possibilities. In a single neuron that is included in the control network, the direction of synaptic weight changes for informational inputs can be designated by the sign of an initiating signal (Figure 47a, b). The situation may be even more interesting when the signal that acts as an initiating stimulus arrives in the network with bolt initiating and informational components (see Figure 46). If the initiating component arrives several milliseconds later than the informational component (we shall see in Chapter 10 that this possibility is realized in some structures), then the latter will be treated by the neuron like any other informational channel, and changes in its synapses can occur. For the system, this means that the response evoked by the informational component of the initiating signal can be perpetually and precisely tuned by the learning process so that the response optimally reduces the duration of influence of the initiating stimulus itself in the absence of *any* additional informational context. Moreover, in this case, the direction of the neuronal response—whether excitation or inhibition—to the informational component of the initiating signal can be used to define the direction of the changes in synapses of other informational channels that are necessary for removing the influence of the initiating stimulus. This is a very significant improvement that can dramatically accelerate the learning process. Let us consider several additional possible situations. In the first scenario, an initiating signal excites the neuron, indicating the presence of an error. The informational component of the initiating signal can be excitatory or inhibitory (Figure 47c, d), and its sign determines the direction of changes in the synapses of other informational channels. In the second scenario, which is the opposite of the first (and theoretically possible), the initiating component is inhibitory, while its informational component may be excitatory or inhibitory (Figure 47e, f), and its sign also determines the direction of changes in the synapses of other informational channels. Obviously, the existence of the combinations described above is very important for the execution of proper control, because the correct direction of changes in informational synapses for different portions of the neural network, whose interactive relationships during a control task may be antagonistic, might therefore be different.

It is understandable that many neural circuits can learn according to a classical conditioning scheme by utilizing different excitatory and inhibitory combinations of initiating and informational signal components. However, the general principle remains the same in all cases: the initiating signal evoked by an unconditioned stimulus has to be minimized by the end of the learning procedure by utilizing the available informational context (conditioned stimulus) to compute the correct output.

It is well known that the most efficacious conditioning paradigm involves simultaneous conditioning, in which the CS terminates at the end rather than at the beginning of the US, while trace conditioning is the least efficacious paradigm. The reason for this is obvious: The presence of afferent signals consisting of an informational context that contains a particular CS is necessary for this type of learning when the CS is followed by an US consisting of an initiating signal. In delayed conditioning, trace processes may be responsible for providing the system with the necessary informa-

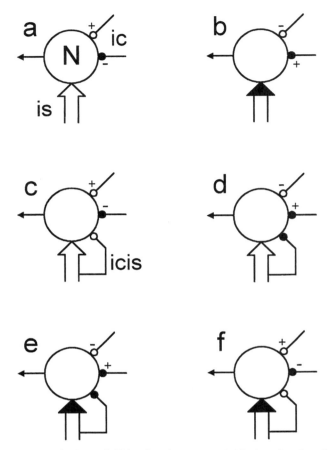

Figure 47—Theoretical possibilities for the way an initiating signal or its informational component can direct a change in the synaptic weights of informational synapses. N, neuron; icis, informational component of initiating signal; double arrow, excitatory initiating signal; double arrow with black head, inhibitory initiating signal. Excitatory and inhibitory informational inputs are designated by white and black circles, respectively. + and − designate, respectively, an increase and a decrease of the synaptic weight in corresponding informational synapses. In a and b, the direction of synaptic weight change in informational synapses is dictated by the sign of the initiating signal. In c–f, the direction of the synaptic weight changes in informational synapses is dictated by the sign of the informational component of the initiating signal. f is a theoretically possible but very unlikely situation.

tional context, which, obviously, has lower intensity than in the case of simultaneous conditioning. Higher centers have to provide the system with the informational context in the case of trace conditioning, in order to fill the gap between the end of the conditioned stimulus and the beginning of the unconditioned stimulus.

7

The Hierarchy of Neural Control Systems

Hierarchy is a widely used term in neurobiology. Its use is most often based on an intuitive understanding of the hierarchy concept and implies the subordination of one control level to another. A military hierarchy is a common example of herarchical relationships. A general gives orders to his officers who, in turn, give orders to their soldiers. A variety of other examples of hierarchies can be provided. However, one has to identify the very nature of a hierarchy in order to proceed with any theoretical reasoning. Otherwise, it is impossible to answer the following very important questions: Why are complex control systems hierarchical? And what is the major functional advantage of a hierarchy in a control system?

To develop a mathematically precise definition of a hierarchy is a very complex undertaking. This is why scientists usually offer a working definition suitable for a specific purpose. For instance, R. Hecht-Nielsen (1990) defined hierarchical neural networks as "networks with sparse and localized connectivity between layers."

For our purposes, we will focus on the following features of a hierarchy. When moving from lower to higher levels, generalization or abstraction of parameters within the hierarchy occurs. The parameters of higher levels change less frequently than the parameters of lower levels. This is why in the case of motor control, a simple command from a higher level can initiate a rather complex automatism controlled by a lower level. In general, the hierarchy present within a control system has to be considered to be a consequence of an object state hierarchy (Baev and Shimansky, 1992). For instance, a multi-joint limb is a hierarchical object. Several major principles, according to which hierarchies are built in biological neural networks, can be established if one analyzes how OCSs of different levels must be connected, taking into account that each OCS is a learning system containing a model of object behavior.

Let us recall how systems initiating lower automatisms work. The concept of a command neuron was described in Section 1.1.2 for the case of locomo-

Biological Neural Networks
Konstantin V. Baev
© 1998 Birkhäuser Boston

tion and scratching. Numerous other command areas have also been demonstrated experimentally in vertebrates. Brain stem and spinal cord *command areas* are responsible for organized acts such as eating, gnawing, licking, lapping, sexual acts, aggression, and many other types of inborn behaviors. These command areas also activate corresponding automatisms that are capable of coordinating the many subcomponents of behavioral acts, taking into consideration the sensory feedback arising from each of the various subcomponents.

Therefore, in the case of the initiation of lower automatisms by a command system, a *trajectory* at the lower level corresponds to a *point* within a given system state at this higher level of command neurons (Figure 48). Suppose the body parts are moving along specific trajectories in a time interval t_0–t_1. After time moment t_1, the system switches to movements along another set of trajectories. They may be *closed* trajectories, meaning that the system performs cyclic movements. It is obvious that each initiating system has to "jump" from a state that sets the parameter P_{1j} for the lower trajec-

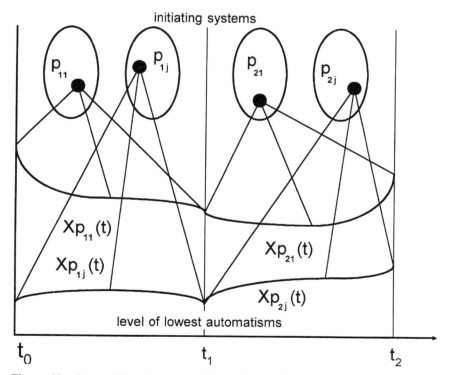

Figure 48—Interrelation between different hierarchical levels. p, parameters that the higher level sets for the lower one; x(t), trajectory of movement. See text for explanations.

tory to a state that sets the parameter P_{2j}, in order to switch to another trajectory. In turn, the command system level can be a controlled object for a correspondingly higher level. In this case, a *single cycle* at the command level corresponds to a *family* of trajectories at a lower level.

Thus, the higher level "imagines" the behavior of the object in a manner that is different from the way the lower level does. A *dynamic* model of the lower level predicts the state of the object in the next moment in time. The higher level model thus predicts in *what state* and *when* the transition of the object to its next state will take place—that is, it is a cause-effect model of the controlled object that predicts a series of object states. This includes the coding of probability distributions for the time intervals after which these states must appear.

Figure 49a illustrates what signal types are sent by one hierarchical level to another. It is easily understood that the controlled object sends to its controlling system—to lower OCS—the same types of signals that the lower OCS sends to a higher OCS. This is because the lower OCS is a controlled object for the higher OCS. Other routes for initiating; mismatch, and informational signals also exist in hierarchical biological neural networks. Some of them are shown in Figure 49b. The advantage of these connections is obvious. For instance, the control abilities of a lower level are significantly

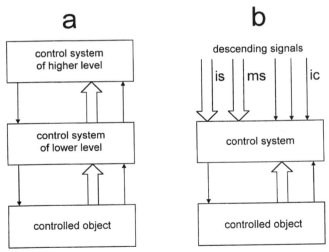

Figure 49—Types of signals that different hierarchical levels exchange with each other. a, control system of lower level is a controlled object for the higher one, and it sends to the higher level the same types of signals that it receives from its controlled object; b, higher level control systems may provide lower level control system with informational context and different types of initiating signals; ms, mismatch signal. Other designations as in Fig. 44.

extended if it receives mismatch signals from higher levels, because new minimization criteria are added. The higher level can also provide the lower level with an informational context, of a type that does not exist at the lower level. For instance, vestibular information for the spinal cord, visual and acoustic information for the cerebellum, etc.

The above examples serve to illustrate such concepts as the descent of an automatism. Let us consider a simple example. Suppose a higher level learns the salient features of a given situation earlier than a lower level, and therefore "knows" what to do in a new situation. This higher level sends initiating and informational signals to the lower level. The latter does what it can, and learns to minimize the initiating signal by using the available informational context from this higher level, and from other sources, according to the principle of learning described in the previous section. As a result of this process, the new afferent signals that are now available to the lower level will also be included in the model of this lower level. The higher level can remove the initiating signal sent to the lower level when this process of automatism descent is complete.

Obviously, there should be a match between the control level and its detectors, because the latter have to properly describe the corresponding space state coordinates. For instance, a command system initiating a locomotor automatism can operate with such a parameter as the intensity of locomotion, a parameter that does not exist at the level of a CPG and is the result of hierarchical generalization of lower level parameters. It is possible to consider the complexity and sophistication of detectors of sensory information from the same hierarchical perspective. One can formalize any "receptor" the following way (Figure 50a). It fires when the stored model does not coincide with the controlled object afferent signals; the receptor threshold thus plays the role of the model. This signal arrives at the next control level in the network hierarchy, which in turn is the controlled object for a higher control level, etc. (Figure 50b). From a formal point of view, it is possible to consider any descending influence on any system transferring ascending afferent flow as a *control influence* of the control system on its controlled object. In such a system, a lower detector is fixed upon a particular feature when a corresponding descending initiating signal arrives. On the other hand, a mismatch signal will go higher and higher until it is caught by a competent level capable of removing this signal.

The architecture of a primary cortical detector is shown in Figure 51a. Mismatch signals reach the cortex through nonspecific thalamic nuclei. Obviously, the portions of the reticular formation that project to nonspecific thalamic nuclei may be considered to be a part of an error distribution system. It is also well known that the cortical level can control ascending afferent systems, for instance, spinal sensory ascending systems. Clearly, secondary, tertiary, etc. cortical detectors can be created within the cortex itself according to the principles described above (Figure 51b). As a result

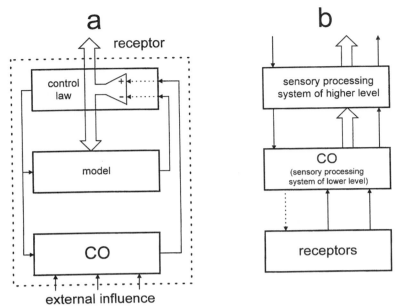

Figure 50—Formalized view of a sensory processing network. a, sensory receptor; CO, controlled object: receptor metabolic processes, excitability, etc; b, interrelation between different hierarchical levels of a sensory system. Sensitivity of some receptors may be centrally controlled.

of this process, a hierarchy of sensory cortical detectors is created. It is noteworthy that higher sensory levels can send initiating and informational signals to lower detectors and, consequently, tune them for any desirable feature because new minimization criteria become available to the lower levels.

Therefore, the principle of hierarchy significantly facilitates the resolution of any complex control problem for the whole control system. In a hierarchical system, each level becomes responsible for its own automatisms and computational problems for higher levels may not be more difficult than those of lower levels. But any movement toward higher levels leads to new computational problems and has to be accompanied by the development of specific computational, learning, and memory mechanisms, because working space becomes more and more discrete and the working time interval more and more extended.

In the discourse to follow, several different mammalian brain systems and their hierarchical levels will be analyzed according to the concepts described above. This will be done by using a rather simple algorithm: For each level, we will attempt to identify both functional subsystems—the controller and

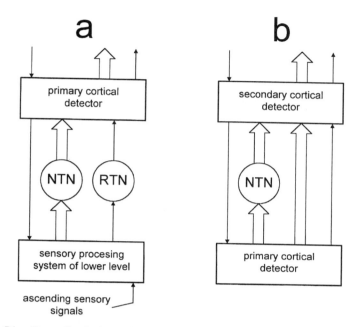

Figure 51—Formalized view of cortical detectors. a, primary cortical detector; b, interrelation between primary and secondary cortical detectors; NTN, RTN, nonspecific and relay thalamic nuclei, respectively. Mismatch signal of primary cortical detector can also reach secondary cortical detector through nonspecific thalamic nuclei.

the model—as well as the controlled object, and attempt to determine the various types of afferent signals—informational and initiating. Special attention will be paid to error distribution systems. This approach will demonstrate that, given the theoretical concepts developed up to this point, we can draw several very important conclusions about the organization of higher levels in the nervous system. Surprisingly, these conclusions can be reached without a detailed knowledge of the computational abilities of the underlying neural networks.

8

Application of the Concept of Optimal Control Systems to Inborn Motor Automatisms in Various Animal Species

We can now provide a natural and simple explanation for the experimental facts described in Chapter 1. We will focus on the initiation of CPGs and their organization in various animal species.

8.1 The Principle of Motor Automatism Initiation

Let us confine ourselves to considering only the initiation of locomotion during avoidance behavior. One source of initiating signals is the controlled object itself. Painful signals are a typical example of initiating signals of peripheral origin. Tactile, temperature, and other types of contact receptors, as well as vestibular and distant receptors, are all sources of informational context.

The evolution of motor automatism initiation systems consisted of the creation of neural networks able to compute a controlling output that minimized the external initiating signal—more precisely, the *potential* reception of any initiating signal. Such evolutionary improvements proceeded according to the classical conditioning schema described in Section 5.2.4 and can be designated as *evolutionary learning*.

Let us consider several consecutive schemata that demonstrate the increasing complexity of initiating systems in evolution.

1. The neuronal network responsible for motor control receives only the initiating signal, e.g., a painful stimulation evoked by contact with an enemy. Evolutionary improvement for such a network is limited. A decision can be made by which the network gains the ability to de-

Biological Neural Networks
Konstantin V. Baev
© 1998 Birkhäuser Boston

crease the duration of the initiating signal influence by performing a phasic movement of part of the body or the whole body away from the irritant. The system cannot avoid the onset of the irritant influence because the initiating signal is the only source of information. The appearance of networks that are capable of generating a sequence of stereotyped movements (locomotion, shaking, etc.) could be the next step in the improvement of such a system. It is possible that, in those cases in which the initiating signal has a tonic character—for instance, a continuous dangerous influence from the environment—to depart from the current location may be the only means of escaping from the situation. In such a case, rhythmic locomotor movements lead to an effective decrease in the duration of the initiating influence because the animal moves farther away from the enemy or removes the irritant from its body.

2. In addition to receiving initiating signals, the neural network also receives informational signals, e.g., signals from tactile receptors, antennae, vibrissae, etc. This improvement gives to the system a number of advantages over the first form of the system. The evolutionary improvement can lead to an even greater ability to reduce the influence of initiating signals, and probably even to avoid the initiating painful stimuli themselves when informational signals precede initiating signals. Using information from tactile receptors, a neural network can develop to the point where it is capable of learning to calculate such sophisticated control influences on the object, so as to move it in advance to a state in which the subsequent reception of an initiating signal is minimal or even absent (at the expense of movement farther from the enemy).

3. The next step of neural network complexity is usually accompanied by the appearance of distant receptors such as visual, acoustic, and olfactory receptors. The ability of such a system to decrease a potential initiating signal is even greater than that of the system described above. This more complex system can generally even avoid contact with the enemy or other dangerous influences. For obvious reasons, such initiating networks are located in the regions of the brain that are in close proximity to the abovementioned distant receptors.

It is evident that the same strategy for evolutionary improvement is used in the second and third cases, and consists of utilizing any available informational context to compute an output that minimizes potential initiating signals. The more complex the animal, the more sophisticated the informational context it can analyze to calculate control actions that minimize initiating signals. The informational context can include signals from specific detectors of varying complexity; this provides the network with the potential for creating hierarchical initiating systems.

8.2 Invertebrate Central Pattern Generators from the Perspective of the Optimal Control System

Current ideas about the organization of two invertebrate control pattern generators were described in detail in Section 1.1.3. Let us try to interpret their organization using the OCS concept.

Those examples convincingly demonstrated the absence of a general plan for generator construction in invertebrates. But this does not mean that there is an absence of general principles of construction for such systems. It is possible to easily identify such principles using the OCS concept (Baer and Shimansky 1992). The generators for locomotion in the sea slug and sea angel could functionally be considered to be OCSs using their internal models of controlled objects for producing motor programs, in the absence of peripheral feedback. Let us try to find such models in these neuronal systems.

In the sea slug and sea angel, temporal patterns of swimming movements are quite simple and consist of the alternation of dorsal and ventral shifts of the body and "wings," respectively. Therefore, it is enough for the internal model of the corresponding controlling systems to describe the major dynamic characteristics of the controlled object (such as its inertia and the water viscosity). The physical properties of the controlled object determine the duration of the motor phases during specific activating signals arriving in the muscles. The higher the intensity of the movement performed, the shorter the duration of the motor phases, i.e., the shorter is the corresponding phase of motor activity. It is clear that *wings*, having lesser mass, can perform fast movements more easily. Such regularities in the normal behavior of the controlled object are apparently modeled by the neural networks for these organisms.

Unfortunately, data regarding the afferent control of these two generators are absent, and we cannot consider such important points as interactions between internal models with peripheral afferent signals. Consequently, in our subsequent reasoning, we can only speculate about how afferent signals reach a generator.

Information regarding the spatial and temporal correlations necessary for describing controlled object behavior and for the functioning of internal models is coded in the functional features of neurons, as well as in the structural and functional features of their synaptic connections. If afferent signals reach DSI and VS-B interneurons in *Tritonia,* then chain DSI-C2-VSI-B may be considered to be the internal feedback that forms the basis of a dynamic model of the dorsal flexion of the body during a given controlling influence on dorsal motoneurons. As we have seen earlier, the characteristics of synaptic transmission and the parameters of the A-current in VSI-B play an important role. Similarly, the neuronal circuits involving interneuronal types 7 and 8 therefore model the dynamics of corresponding wing movements during slow swimming in the sea angel. The involvement of type 12 interneurons provides the ability to model fast swimming movement. If we suggest that the generator includes n type 12 neurons, then 2^{n-1} regimes of fast swimming and their corresponding model states are possible.

Thus, the functional essence of mutual inhibition paralleled by delayed excitation, which can be demonstrated in both generators, constitutes the modeling of the dynamics of controlled object behavior. In the case of the simplest generators, the dynamic model of the controlled object can be partially realized by the use of pacemaker neurons (for instance, types 7 and 8 in a sea angel) whose period of rhythmic activity is temporally close to the period of swimming movements. In this case, internal feedback—a model substrate, in this case—is included not only in the generator network structure, but also in single neurons.

It is necessary to stress once again that the activity of a generator cannot be completely synonymous with the functional activity of the model of the controlled object, in spite of the fact that most generator neurons, if not all, are themselves the substrates for the model. From a functional perspective, an OCS calculates control signals according to the current state of the controlled object (according to the state of the internal model in the case of deafferentation). Such a calculation in the system examined above consists of the simple recoding of the model state into controlling signals (for instance, DSI bursts reach motoneurons, on the one hand, and the model on the other hand). Thus, in the above-described generators, the interaction between the functional subsystems realizing the model and the control law occurs in full accordance with the schema presented in Figure 35.

It is necessary to emphasize once more that neural networks, being the basis for the internal model and the optimal control law, cannot be anatomically isolated. Let us recall for a moment the spinal neural networks described in Chapter 4. Even motoneurons, which are usually not considered to be a component of the generator itself, could functionally be considered to be sensory neurons and may therefore be included in the neural network that models the controlled object behavior. By the same token, the presence of electrical connections between type 7 and 8 interneurons and motoneurons allows us to identify the latter as being included in the internal model of the controlled object. At the same time, electrical and chemical synapses *between* interneurons and motoneurons, as well as between motoneurons, could be considered as reflecting a mechanism for coding the optimal control law. Excitatory and inhibitory connections reflect positive and negative correlations between signals, respectively. In the light of the above description, the absence of a general plan for generator construction in invertebrates becomes clearly understandable. Such a plan cannot exist. How could the same plan have been used to build dynamic models for different types of movement in different animals, and their corresponding optimal control laws? According to the OCS concept, control neural networks should be considered as memory—for storing information regarding the optimal control law and the model of the controlled object—from which information can be retrieved under the influence of initiating signals.

8.3 Vertebrate Central Pattern Generators from the Perspective of the Optimal Control System

Now, let us apply the OCS concept to lamprey and *Xenopus* generators. Two features peculiar to the object behavior of these animals during swimming can be readily observed: the first is the alternation of segmental muscle contractions (and body flexions, respectively) on both sides, and the second is the phase lag between movements of neighboring segments. These features of the corresponding motor programs are encoded in the spinal OCS. The realization of the first feature requires the establishment of a negative correlation between the neuronal elements of the two sides of each segment, so that they work out of phase with each other. The basis for this negative correlation could only be inhibition; therefore, cells inhibiting contralateral neurons—including motoneurons—must exist. This has been precisely demonstrated in numerous experiments (see Section 1.1.3).

Phase lag calculations belong to functions realized most easily on the basis of neural networks. Temporal and spatial summation of influences coming from neighboring spinal cord segments (these influences must decrease with increases in the distance between these parts) is the most simple means of realizing such functions. The corresponding neuronal connections have also been demonstrated in lamprey and frog embryos.

The example of lower invertebrate swimming demonstrates clearly that the model of controlled object behavior cannot simply be reduced to a model that calculates only the period of rhythmic processes. Nevertheless, the development of models that simulate rhythm generation was precisely the goal of numerous previous theoretical works. As we saw in Section 1.1.3, parameters of rhythm generation can easily be set by a neural network, or even by a single neuron. This very schema is used in the lamprey generator: spinal interneurons and motoneurons have pacemaker properties, and the major role of the network consists of defining the necessary correlations among the activities of numerous nerve cells.

Thus, the OCS concept makes it possible to examine the generator systems of different animal species using a few general principles. The major difference between generator systems in lower and higher animal species is that more complex internal models and control laws are encoded in the latter, and, consequently, the latter OCSs perform more complex afferent information processing (Baer and Shimansky 1992).

We can now provide a natural explanation for the widely used concept of *half-centers*. Correlations between activities of single neurons are encoded by the sign (excitatory or inhibitory) and the weight of their synaptic connections. The presence of two cell groups working out of phase is usually observed in generator systems whose effectors may be divided into two antagonistic groups. The activity of these effectors leads to activation of the corresponding groups of peripheral receptors, whose activity is modeled in the OCS. It is obvious that the existence of alternately active groups of cells called *half-centers* reflects the presence of a negative spatiotemporal correlation between the activities of the two groups of cells during the generation of motor programs. The appearance of cells whose activity pattern contradicts the half-center hypothesis is natural for systems controlling complex movements that require complex coordinated effector patterns, in comparison with simple alternation. It was mentioned in Section 1.1.3 that the phases of the switching on and off of single neuronal activities are almost evenly distributed in the scratch and locomotor cycles of cats.

Convergence and divergence of nerve terminals in the nervous system can be considered to be the basis for the encoding of numerous correlations between the activities of different neurons and receptors. A typical example reflecting the presence of positive correlations encoded in the neural network is the organization of synaptic terminals of Ia fibers in the spinal cord. A single Ia fiber makes contact with almost all motoneurons innervating the same muscle. The convergence of afferents of different modalities on a single neuron could also be easily explained from this perspective.

8.4 The Phenomenon of Entrainment of Central Rhythms

In the previous portions of this section, we paid little attention to such important issues as the application of the principles of afferent control (described in Section 1.1.4 and Chapters 3 and 4) to the simpler generators in invertebrates and lower vertebrates. The generalization of the OCS concept was based primarily on the well-known fact that a CPG produces a motor program similar to that of an intact animal without the aid of peripheral feedback, and it was shown that a CPG deprived of peripheral afferent signals produces motor programs using only the internal sensory model.

The phenomenon of central rhythm entrainment deserves special attention because the existence of this phenomenon is a serious argument in favor of a positive response to the question of whether or not it is possible to apply the OCS concept to simple systems. The phenomenon of entrainment was found in all CPGs in which the influence of rhythmic afferent signals on locomotor activity was investigated (see, for instance, Sillar and Scorupsky 1986; Sillar et al 1986; Grillner et al 1982). It is not expedient to describe all these results in detail. So, let us confine ourselves to making reference to a few experimental facts only. In the crayfish, rhythmic stretching of the thoracic-coxal muscle receptor organ entrains a central locomotor rhythm produced by the thoracic ganglia in which remotor and promotor motoneurons of the leg are alternately discharged. The entrainment of central rhythm by mechanical cyclic movement of the spinal cord was found in experiments on isolated lamprey spinal cord. Intraspinal mechanoreceptors—the so-called edge cells—are responsible for this effect. Using the OCS concept, peripheral and model afferent flows interact on a parity basis. This is completely true in the case of the lamprey. Both model (centrally originated signals) and peripheral (signals from the edge cells) spike activity can modulate the generator rhythm in the lamprey. Central signals and signals from the edge cells have equal ability to influence a postsynaptic neuron. The postsynaptic neuron does not see any difference between them and can use both signals without bias or preference based on their origin. In the case of higher vertebrates, functional mutual inhibitory interactions based primarily on presynaptic inhibition are an important attribute of parity interactions. In the lamprey—a lower vertebrate species with a nervous system that is much simpler than that of higher vertebrates—this mutual functional inhibitory interaction can be based, for instance, on the nonlinear properties that govern their summation: a signal that arrives first has the greatest influence on the system.

It is obvious that the phase dependence of generator reordering under the influence of afferent flow must be the basis for the phenomenon of entrainment. Moreover, as was shown in Chapter 3, reordering plots must include those portions having positive slopes and corresponding NR points. The presence of the latter explains how synchronization between the generator and cyclic afferent flow takes place.

Let us consider swimming movements. There is usually a certain temporal delay between the emanation of a central command and the corresponding movement. As a result, the mechanical phase lag is greater than the neuronal phase lag (Grillner and Kashin 1976). It follows from the functional system feature of optimality that during undisturbed swimming, the maximal afferent signals should arrive in their corresponding segments at time moments corresponding to the phasic vicinities of NR points. Therefore, in different animals, the relationship between mechanical and neuronal phase lags must be tightly coupled with the organization of body mechanoreceptors and their central terminals in the spinal cord.

A mechanism similar to that described above is obviously the basis for synchronization of neurons that have pacemaker properties, and is included in the lamprey locomotor generator. Examples of rhythm changes under the influence of de- and hyperpolarizing current pulses are shown in Figure 52. Portions of the positive slopes intersecting the level corresponding to the absence of reordering are easily identified here.

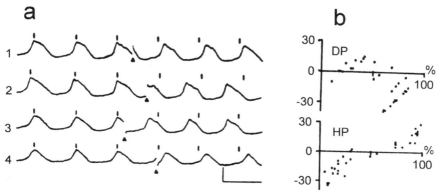

Figure 52—Effect of de- and hyperpolarizing current pulses on TTX-resistant oscillations of membrane potential (according to Grillner et al 1986). a, positive (1, 2: 1 pA, 20 ms) and negative (3, 4: 2 pA, 20 ms) current pulses evoking rhythm reordering. Vertical bars designate normal appearance of maxima of membrane potential oscillations. Calibration: 10 mV, 4 sec.; b, reordering of cycle duration during application of depolarizing (DP) and hyperpolarizing (HP) pulses of current.

While analyzing the afferent role in automatic movement control, it is necessary to again focus our attention on the fact that some authors present unlikely explanations for the similarity between motor programs generated before and after deafferentation (Arshavsky et al 1985a). They conclude that the sea angels swimming motor pattern does not require afferent correction because the external conditions for wing movements do not usually change; this is a major difference between swimming and terrestrial locomotion, which requires peripheral feedback for adaptation to external conditions. This explanation is highly unlikely to apply in the case of an animal as complex as the sea angel. However, it cannot be completely ruled out for very simple generator systems (see Conclusion).

8.5 A Generator is a Learning System!

It follows from the OCS concept that the internal model must describe the controlled object behavior as precisely as possible. If the model is not adequate (signals coming from the periphery are not corresponding to the model ones), the control system cannot function correctly. Let us recollect, for instance, how difficult it is for a man to use his limb after a long period of limb immobilization. In this connection, any long-term change or progressive alteration in the characteristics of the controlled object requires that corresponding changes be made in the model and, consequently, the OCS must be able to adjust the model to the object, i.e., to learn. There are many examples of short- and long-lasting changes in the controlled object. Rapid compensation for an abrupt change in the weight of the controlled object (the weight of shoes, locomotion with prey caught by an animal, etc.) and learning to overcome an obstacle encountered in each locomotor cycle are examples of rapid but short-lasting changes. It is possible to observe the process of learning to overstep an obstacle even in decerebrated animals (Lou and Bloedel 1988). Long-lasting changes take place during organismal growth, when the geometric size of the organism's moving parts—their elastic properties, and other features—are changing. Such changes are most rapid and intense at the early stages of ontogenesis.

It is obvious that the necessity for short- and long-lasting corrections of the model requires corresponding memory mechanisms to be present in the OCS. The algorithm of such learning based on the method of trial and error, as well as the signals initiating such learning processes, was considered in Chapter 6. Here we will only consider some experimental facts testifying to the ability of an OCS to learn.

Many animals are born with well-organized motor functions whose development takes place during embryogenesis. This so-called embryonic motility already appears at the early stages of embryogenesis and undergoes several stages. Perhaps the most detailed investigation was performed on the chick embryo (Hamburger 1963; Hamburger and Oppenheim 1967). Such embryonic movements are characterized by random twitches that become progressively more complex during development. These movements evoke afferent signals necessary for correct tuning of the internal model of the controlled object by the method of trial and error.

There is a massive death of spinal motoneurons (probably also interneurons) from the sixth to nineth day of incubation (Hamburger 1975). This could be explained by the fact that these cells might not have been incorporated into the evolving model substrates. Such an explanation is in good agreement with the decline in cell death noted following pharmacologic treatment (Pittman and Oppenheim 1979) of chick embryos that receive daily in ovo injections of cholinergic blockers. During this treatment, the OCS does not receive information about object behavior and, consequently, it receives no information about any mismatch between the latter and its internal model, i.e., it receives no signals about errors initiating the learning process.

The central modulation of PAD is already present at the early stages of embryogenesis. At the late stages, it is more prominent (Chub and Baev 1991). The intensity of antidromic discharges appearing at the top of PAD waves in dorsal roots is comparable with ventral root activity. Such dynamics of presynaptic inhibition during embryogenesis agree with the concept of learning. After the construction of a rough model at the early stage of embryogenesis, the stage of fine-tuning begins.

As mentioned above, the generators could be considered to be memory systems that store information about the optimal control law and the dynamics of the controlled object. Different states of this system correspond to its different regimes. Memory systems are formed during evolution and reproduced in ontogenesis with the help of genetic mechanisms and learning processes. The features of learning processes are obviously critical components in the formation of generator memory systems. It is impossible to imagine that the whole structure of synaptic connections, including their weights, could be encoded in DNA, because the information capacity of the latter is insufficient. Moreover, it would also be necessary to encode the program of changes in synaptic weights that occur during growth in ontogenesis, and small deviations in development, which could be easily compensated for by learning, might lead to animal death. That is why nature presumably chose the foregoing alternative strategy: DNA encodes the system of rules of construction and development of basic neuronal OCSs, and all remaining aspects of development are left to learning.

Holographic properties of the brain are most prominent in vertebrates. As was mentioned in Section 1.1.3, a comparatively small portion of the spinal cord can produce activity that contains the principal features of locomotor pattern, i.e., a readout of information encoded in the network is performed under the influence of an initiating signal. This feature is widely used at present in experiments *in vitro*. Moreover, even a small part of the gray matter of one spinal cord segment can produce a locomotor-like pattern. This was demonstrated in experiments on the spinal cord of chick embryos (Baev and Chub 1989; Chub and Baev 1991).

In Section 1.1.4, it was mentioned that the central modulation of PAD reflects the work of the generator. In this connection, it is possible to observe generator activity by recording not only efferent signals (as is usually performed in experiments), but also by recording signals in dorsal roots. PAD modulation permits the evaluation of the generator activity even after anatomical separation of motoneurons from the generator itself. Experiments have shown that the isolated dorsal horn can rhythmically modulate PAD. Moreover, this rhythmic activity is NMDA-sensitive, i.e., it is enhanced after the administration of this drug (NMDA is a substance that specifically activates the locomotor generator). The rhythmic activity in ventral roots is only observed after sectioning the ventral region of the spinal cord that includes the ventral portion of the dorsal horn. These facts are easily explained from the perspective of the existence of an internal model. Dorsal horn interneurons are the target for numerous afferents of different modalities, and parity interactions between the model and peripheral signals take place here.

At first sight, these facts contradict contemporary concepts of the generator. On the basis of numerous data obtained with microelectrode techniques, a hypothesis appeared that proposed that locomotor and scratching generators are constructed utilizing spinal interneurons located in the lateral regions of the intermediate zone and ventral horn (see Chapter 1). A maximal number of rhythmically active neurons has been identified in these regions during fictitious scratching and locomotion. In experiments *in ovo*, it was also found that a maximal level of multineuronal activity was recorded during spontaneous movements from the ventral portions of the spinal cord (Ripley and Provine 1972). Experiments with the 2-deoxyglucose method (Viala et al 1988) have also revealed that a large quantity of this glucose derivative accumulates in the intermediate zone of the cord gray matter during locomotor generator activity in the rabbit.

This contradiction can easily be explained. Extracellular electrode recordings are better performed on larger, rather than smaller, cells. It is possible that there are significantly more rhythmically active cells in the dorsal horn than were actually identified experimentally. Furthermore, despite the fact that the generator subsystems—the internal model of the controlled object and the subsystem calculating the law of optimal control—cannot be separated anatomically, their corresponding neuronal structures are nevertheless primarily localized in the dorsal and ventral regions of cord gray matter.

9

The Stretch-Reflex System

Let us examine the lowest motor control level for which the neural structure is delimited by the major identified neurons of the spinal cord (i.e., α- and γ-motoneurons, Ia interneurons, and Renshaw cells) and by extrafusal and intrafusal muscle fibers. The experimental data on the structure of synaptic connections between cells of these types and the characteristics of their activation during movements are well known (see, e.g., Granit 1970). Therefore, we will briefly enumerate the most important of these facts without references.

The pattern of connections between cells of the abovementioned types is shown in Figure 53. Alpha-motoneurons and Ia interneurons are obviously classifiable as sensory neurons according to their functional properties. They can be activated both by peripheral afferent signals and by signals from central neurons. Signals coming to a muscle spindle from γ-motoneurons lower the activation threshold of its spindle receptors. These signals can be regarded functionally as exciting the corresponding flexor and extensor Ia afferents (Ia af$_f$ and Ia af$_e$). The activity in α-motoneurons of a given muscle can be considered as functionally inhibiting these muscle spindle receptors. In addition, the action of an external effort (F_f and F_e) directed toward stretching a given muscle has a functionally excitatory character on the spindle afferents.

The simplicity of the above-described construction and the well-known role of negative feedback in control mechanisms provide an illusory basis for the hypothesizing of a theory, though applicable only to this motor control subsystem. Many attempts have been made along this line. Among the first ideas was the hypothesis of Merthon (1953)—the hypothesis of *servocontrol*—which was apparently the most popular. According to this hypothesis, γ-efferent signals are designated as *servo* signals and set the muscle lengths conforming to the aiming state in a given postural motor reaction. Alpha-motoneurons receive information about them through Ia

Biological Neural Networks
Konstantin V. Baev
© 1998 Birkhäuser Boston

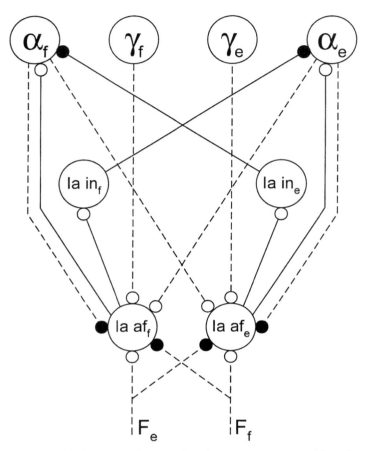

Figure 53—Simplified pattern of connections between neurons, and functional connections in the system of stretch reflex. Functional connections are shown by dotted lines. Ia in_f and Ia in_e, Ia flexor and extensor interneurons, respectively; F_f and F_e, force applied by flexor and extensor muscles, respectively. Ia af_f and Ia af_e, Ia afferents of flexors and extensors, respectively. See text for explanation.

afferent feedback and distribute motor control commands directly to the "executive organs." It is assumed that the stretch-reflex system must stabilize the necessary value of muscle length. However, the phenomenon of a-γ-coactivation, discovered later, refuted this hypothesis.

The latest and most carefully elaborated hypothesis for explaining the interaction between central and peripheral influences in the stretch-reflex system is, apparently, the *equilibrium point* hypothesis proposed by Feldman (1979). The main idea is that central signals set the threshold of stretch-reflex initiation rather than a discrete muscle length value. In other words, central influence determines the equilibrium state to which the system "muscles—external load" should move.

The equilibrium point hypothesis is very close to the idea described above about the interaction between different levels of a hierarchical OCS. However, this hypothesis limits the set of possible movements by postural reactions. Furthermore, as has been experimentally demonstrated, the condition of a muscle depends not only on central signals arriving at the current point in time, but also on its previous history of movement (Kostyukov 1987) as well as on the previous history of activation of a- and γ-efferents (Kostyukov and Cherkassky 1992). In other words, a muscle has a pronounced hysteretic property. Hence, the equilibrium point hypothesis over simplifies the real system.

What conclusions regarding the stretch-reflex neural system can be drawn according to the OCS concept? Let us try to represent this system on the functional level in the form of an OCS containing a model of its controlled object (Baev and Shimansky 1992).

The function comprising the "controlled object state—sensory signal" is encoded in the lowest motor control level by the pattern of muscle attachments to the skeleton of a given limb. A control law is encoded in the same way: a change in the controlled object state, as a function of the force developed due to the activation of a given muscle, is determined by this muscle's attachment to the skeleton of its corresponding limb. It is important that a muscle be regarded as an element of the motor control system and, at the same time, as part of the controlled object—a mechanical system with a certain mass. Given the pattern of interaction between γ-efferent influences on muscle spindles and the mechanical effects of stretching them—a factor for calculating mismatch signals—it is logical to assume that Ia afferents convey the mismatch signals about mismatches between the information produced by a certain internal model and the result of filtering information in the controlled object. Like any mismatch signal, they have to be minimized by the control system. On the other hand, these mismatch signals are sensory signals reflecting the magnitude and the speed of muscle stretching.

Alpha-motoneurons can be excited by central signals and/or by Ia afferent signals. The major difficulty in understanding the functional meaning of central signals that activate a-motoneurons is that they are control signals as well as model signals that should match the corresponding Ia afferent inflow. After deafferentation, the ability to perform postural motor reactions is partially preserved, while γ-efferent influences become ineffective. Consequently, the spinal OCS together with higher nervous system levels can perform such tasks without these efferents. At the same time, if the central influences on a- and γ-motoneurons are absent in the face of intact peripheral afferentation, the stretch-reflex system will be able to work properly in an autonomous regime.

The principle difference between the two situations described above is the origin of initiation of the control system in question. In the first case, it is an effector signal generated by upper motor control levels, while in the second case it is an external influence acting on the limb. Thus, one should suppose that central commands—according to which γ-motoneurons are activated—also act on the internal model of the controlled object to generate an influence on α-motoneurons that models Ia afferent inflow. This, apparently, is the functional significance of the α-γ-coactivation phenomenon.

Let us analyze the case in which there is no extrafusal (α-efferent) control activity. In other words, there are no signals for initiating controlled object motion. If there is an external force directed toward stretching a muscle, such initiating activity cannot be zero. Consequently, the invariant of controlled object behavior corresponding to the internal model of the object in the stretch-reflex system can be described as an inertial dissipative (i.e., with friction) mechanical system. In other words, this controlled object description corresponds to an invariance of potential energy. Thus, all the spatial positions of the controlled object in this particular case are states of neutral equilibrium, and the current coordinates of the object are parameters of its internal model.

Such an object can be regarded functionally as memory, into which information is recorded by muscle efforts according to a differential encoding principle. The signals initiating a change in memory contents are α-efferent signals. The hysteretic property of a muscle endows it with an ability to memorize the previous history of its activation. Therefore, the differential encoding principle should be, and actually is, utilized in generating motor commands. The activation of a motor unit usually begins with a high-frequency bundle of spikes, then the spike frequency decreases sharply to the value necessary for keeping the motor unit on the upper branch of the hysteresis curve. The local negative feedback loop through Renshaw cells and the additional local negative feedback within the motoneuron itself—a condition leading to significant trailing hyperpolarization, a characteristic property of the motoneuron—are actually the substrate of the internal model of the muscle as a controlled object.

The stretch-reflex system alone obviously cannot provide adequate control of rhythmic movements. Therefore, higher-level subsystems in the spinal OCS must be involved in this control task. The stretch-reflex system is used in such situations as a low-level OCS subsystem that is controlled by setting an aiming state of the controlled object (the limb) encoded by γ-efferent activity. As the aiming velocity of the limb is not zero (rhythmic movement), both static and dynamic γ-motoneurons should be involved. In order to facilitate the matching of model and peripheral inflows even during an active muscle contraction, the spinal OCS has to intensively activate γ-efferents simultaneously with the activation of the corresponding α-motoneurons. Such an α-γ-coactivation phenomenon has been observed many times in locomotion (Severin 1970; Perret 1976; Sjostrom and Zangger 1976) and scratching (Feldman et al 1977) experiments. It should be noted that there is no agreement on the explanation of this phenomenon, mostly because of significant discrepancies in experimental results. There are apparently several reasons for such discrepancies.

Most of the experiments were performed by recording electrical activity from single fibers, for which it seems very difficult to observe a common regularity. In addition, the methods for recording γ-efferent activity are very complex and difficult. Taking these obstacles into account, some experimenters (see, e.g., Stein 1982) concluded that the use of macrorecording methods is necessary. However, there are many difficulties in this method as well. In order to completely exclude possible artifacts—depending on the specific conditions under which the experiments were conducted—and to observe what really takes place, the experimenters significantly increased the complexity of the method of recording neuronal activity from a freely moving animal with an intact brain, so that the higher motor control levels were involved. The complexity of such a preparation led to a corresponding complication of the entire picture observed. The pattern of the observed activity lost its regularity, which made it difficult to identify a trend in the results.

Despite differences in the opinions of the investigators, many have come to the conclusion that fusimotor activity should be considered to be something like a template on the basis of which motor control is performed. This idea has appeared primarily because variations in γ-efferent activity were, as a rule, significantly less than those of α-efferent activity. This situation, as well as the fact that the conduction speed in the γ-efferent system is significantly less than that in the α-efferent system, is more easily amenable to an explanation based on the hierarchical OCS concept. Indeed an OCS subsystem is organized as an invariant such that it is necessary to alter its parameters as seldom as possible. It is not hard to imagine a memory system whose current state determines the current distribution of γ-efferent activity. This memory system apparently is functionally located in a higher level of the spinal OCS than the stretch-reflex system. On the other hand, control of α-efferent activity can be regarded functionally as information on a parameter of the controlled object model at the level of the stretch-reflex system. Thus, α- and γ-efferent activities are functionally involved in relaying information with the same meaning, about parameter values for the lower control levels in relation to these activities. Gamma-efferent activity is inextricably dependent on the higher level in contrast to α-efferent activity, and therefore the former alters less rapidly than the latter.

10

The Cerebellum

The role of the cerebellum in movement coordination is experimentally well established. The appearance of this structure in evolution was the natural consequence of the necessity for the nervous system to solve the coordination problem. It is obvious that motor control requires the interaction of different OCSs, each of which controls its own body part. An examination of the possible interactions between simple OCSs that are included in a more complex OCS shows that there are two major mechanisms available to the system for solving coordination problems. The first is the anatomical arrangement that results in mutual interconnections between all possible pairs of OCSs. In this case, each OCS creates an internal model of inflow from other OCSs, because this inflow is a source of its afferent flow. Obviously, such an awkward construction can be relatively efficient in simple systems consisting of small numbers of OCSs. The second mechanism is the creation of one coordinating central dispatcher which receives information from all OCSs and other sources, processes it, and sends corresponding commands back to each OCS. This dispatcher accumulates knowledge regarding spatiotemporal correlations between afferent signals and the optimal control signals that should be sent to each OCS in a particular situation. In previous articles (Baev and Shimansky 1992; Baev 1994), the cerebellum was proposed as the central dispatcher created by nature and elaborated in the evolution of vertebrates.

10.1 The Semantics of Cerebellar Inputs

The conceptualization of the cerebellar coordination function can easily be accomplished by utilizing the semantics of its afferent inputs. These semantics are clearly seen in the example of the activity of the spinocerebellar loop during locomotion and scratching. Such data were primarily obtained by a group of scientists in Moscow and were summarized in a comparatively

Biological Neural Networks
Konstantin V. Baev
© 1998 Birkhäuser Boston

recently published monograph (Arshavsky et al 1986a). No qualitatively new data related to this topic appeared after that publication. The reader who wishes to obtain additional details and numerous references to the literature is directed to Arshavsky's work. In addition to the principal results described in this monograph, some other data will be briefly mentioned below. These experimental facts are as follows:

1. The coordination of locomotor and scratching movements in vertebrates is significantly impaired after cerebellectomy. For example, while the ability to perform scratching movements is preserved, the animal is not able to accurately place an affected limb in the region of a discrete body surface area for scratching with precision (Sherrington 1906).

2. The dorsal spinocerebellar tract (DSCT) is *rhythmically* active during the acts of locomotion and scratching, but this rhythmic activity disappears after deafferentation or pharmacological neuromuscular paralysis when fictive movement is observed.

3. The ventral spinocerebellar tract (VSCT) and the spino-reticulo-cerebellar tract (SRCT) are rhythmically active during both real and fictive rhythmic motor behavior.

4. The spino-olivocerebellar tract has a low level of activity and low modulation during fictive and real unperturbed movement. During locomotion, complex spikes are observed only in response to an unexpected perturbation of movement. Moreover, the spino-olivo-cerebellar pathways that carry information from tactile receptors are gated during active movements so that only non-self-generated inputs are successful in evoking complex spikes (Andersson and Armstrong 1987; Bloedel and Lou 1987; Lidierth and Apps 1990).

5. Descending rubro-, reticulo-, and vestibulospinal tracts are rhythmically active during both real and fictive locomotion and scratching. This rhythmic activity practically disappears after cerebellectomy.

6. When an animal learns to overstep an obstacle placed in its path while the animal is in active locomotion, the complex spikes that occur as a result of perturbation of movement have maximal intensity at the beginning of the learning process, when the limb contacts the obstacle, and minimal intensity at the end of the learning process, i.e., in the absence of contact between the limb and the obstacle during locomotion (Bloedel and Lou 1987). Acoustic signals preceding the appearance of the perturbing bar by several tens of milliseconds can play the role of a conditioning stimulus. This stimulus alone can evoke corresponding changes in parameters of the locomotor cycle.

There are only two afferent inputs to the cerebellum. These are the mossy and climbing fiber systems. Given the experimental results described above, it is possible to conclude that two types of information are relayed to the cerebellum via the mossy fiber system. The first, coming through the VSCT and the SRCT, is information about the state of the model of the corresponding lower OCS, or more exactly, information that is the result of optimal filtering of afferent information in any corresponding lower OCS. The second, coming through the DSCT, can be regarded as actual peripheral afferent information not accounted for by the internal model. Visual, auditory, and vestibular information is also received by the cerebellum via mossy fibers. The cerebellum performs a process of optimal filtering of all the afferent information it receives in order to obtain the most reliable information about the current state of its controlled object, which in this case may be the entire body of the animal and its surrounding environment.

The SOCT, through the climbing fiber system, conveys information to the cerebellum about mismatches that occur at the level of lower OCSs. Mismatch signals appear when the content of model afferent flow does not coincide with peripheral afferent flow. Thus, olivary neurons will be activated during errors of execution of actual movement, during perturbation, and when novel unexpected afferent information is received, such as when a limb comes in contact with an obstacle.

Therefore, mossy fiber input provides input for informational context, and climbing fiber input conveys mismatch information to the cerebellum (Figure 54). It is necessary to note that the cerebellum also receives afferent inputs from other regions of the brain, for instance, from the cerebral cortex. The cerebral cortex sends information to the cerebellum through *both* cerebellar afferent systems, i.e., it sends informational and initiating signals. The system of parallel fibers provides widespread distribution of informational signals across the entire cerebellar cortex.

10.2 How the Cerebellum Learns to Coordinate Movements

A single inferior olivary neuron has about ten climbing fiber collaterals, and each Purkinje cell is connected with only one climbing fiber. Therefore, it is possible to talk about functional cerebellar structure in terms of *modules*. Each module includes Purkinje cells that receive mismatch signals from the same functional part of the controlled object, for instance, a body part. Also included in the module is the neural network that exists proximal to Purkinje cells, i.e., the circuitry that is present from the glomeruli to the Purkinje cells of one module. These proximal circuitries of different modules also partially overlap.

Given the concepts described in Chapter 6, it is possible to conclude that such a module is capable of learning how to compute the necessary output signals that minimize mismatch signals. As mentioned before, such learning

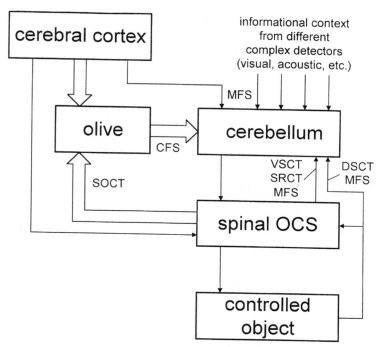

Figure 54—Semantics of cerebellar inputs. SOCT, spino-olivo-cerebellar tract; SRCT, spinoreticulocerebellar tract; VSCT, ventral spinocerebellar tract; DSCT, dorsal spinocerebellar tract; CFS, MFS, climbing and mossy fiber systems, respectively. Designations as in Fig. 40. See text for explanations.

may be effectively organized in the neural network, satisfying several demands:

1. It should have efferent projections to lower OCSs, the activation of which decreases the initiating signal.
2. Corresponding initiating signals should project to such a module.
3. It should receive the corresponding informational context. The richer the informational context, the greater the ability of the system to calculate the necessary control function.

If we recall cerebellar structure, we may conclude that it satisfies the above-described demands. As a matter of fact, functional modules of the cerebellum satisfy these demands. The module includes Purkinje cells that project to the same lower OCS, or to part of it, and receives from it initiating signals through the climbing fiber system. It has been demonstrated experimentally that olivary information supplied to the cerebellum from a specific part of the body is used to influence movements restricted to the same body part (Gibson et al 1987). Purkinje cells of one module receive, via the mossy fiber system, informational context of several different kinds—more exactly, the very different combinations of afferent inputs—each cell receiving its own combination. Patchy organization of the cerebellar cortex has been shown in many experiments (for instance, see Welker et al 1988). Such organization of modules permits the optimal solving of the problem of learning different motor patterns, but it implies the presence of such modules for each body part that is able to perform independent movement. Obviously, these very facts determine the size of the cerebellum: all possible combinations of inputs to different neurons of the same module should be represented in it. Modules are not separated spatially and are intermingled with each other.

The form of dendritic tree of the Purkinje cell is in good agreement with the reasoning made above. It is understandable, even at the intuitive level, that the more complex the dendritic tree, the more complex the functions that can be calculated by the neuron.

However, although in principle the existence of such a learning process is obvious, random searching is very inefficient in this case, and its duration can be excessively long. An interesting solution that accelerates this type of learning could evolve during evolution. Information that is the source of mismatch signals reaches a Purkinje cell through both the mossy and the climbing fiber systems. There is also a very distinct time-lag between the signals of two systems. Mossy fiber information reaches Purkinje cells 10 to 15 ms earlier than climbing fiber information (Eccles et al 1967). What does such an arrangement mean for the entire network system, and how can the system benefit from it? Such a situation has already been discussed in Section 6.2.4. The type of Purkinje cell response to unconditioned mossy fiber input—the informational component of a mismatch signal—might "show" the direction of correct changes in the corresponding synaptic weights of those inputs that are responsible for conditioning, i.e., informational inputs. For instance, if a Purkinje cell responds with *excitation* to the unconditioned component of afferent flow, then the corresponding synaptic weights of those excitatory inputs that are activated by the conditioned component containing an informational context will also be increased (Figure 55). The opposite situation is also possible: if an unconditioned component *inhibits* a Purkinje cell, then the influence of the inhibitory conditioned component will be increased.

The possibility that cerebellar mismatch signals can influence the efficacy of synaptic transmission through those synapses that are responsible for the unconditioned component of afferent flow should not be excluded (see Section 6.2.4). There is nothing to prohibit one from making this conclusion because such synapses are involved in the transmission of the informational

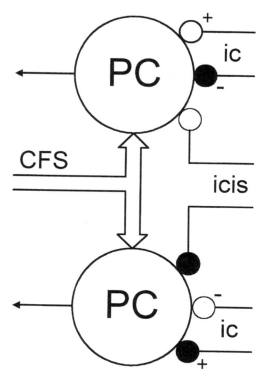

Figure 55—Reciprocal changes in the weights of informational synapses in two Purkinje cells that receive the same initiating signal. PC, Purkinje cell. Other designations as in Figs. 47, 54.

context available to the cerebellum. In this case, this latter mechanism should be capable of persistently and optimally tuning corresponding cerebellar circuitries.

Obviously, a knowledge of the direction of synaptic weight changes accelerates significantly the speed of the learning process. Climbing fibers also project to the cerebellar nuclei, and, consequently, similar learning processes can occur within these nuclei as well. However, the strategy of random search (see Chapter 6) is not completely ruled out; it can be used when the above-described "deterministic" rule is ineffective.

The scheme of learning described above is in good agreement with the experimental results regarding the role of the cerebellum in classical conditioning of the eyelid closure response (see Section 1.2.1). An audible tone—the conditioned stimulus—activates mossy fiber input, while information about the unconditioned stimulus—an air puff to the eye—arrives in the cerebellum through both climbing and mossy fiber systems. As was discussed earlier, when a conditioned response begins to develop, the activity of olivary neurons becomes markedly attenuated, and in trials where the subject reacts with a conditioned response, the olivary neuronal activity evoked by the unconditioned stimulus may be completely absent in trained animals (Sears and Steinmetz 1991). As one can surmise, the conditioned response is computed by the cerebellum *de novo*, which is quite different from the situation in which the conditioned stimulus evokes an unconditioned response.

Both mossy and climbing fiber systems also project from the cerebral cortex to the cerebellum. Therefore, the cerebellum can be provided with new minimization criteria and informational context. This connectivity is a very powerful method of expanding the computational and learning capabilities of the cerebellum. The cerebellum itself is capable of working in rather short time intervals, probably tens or hundreds of milliseconds. This is exactly what the cerebellum does in coordinating movements. It has to work in real time, and simple recoding of input to output has to be used, because there is no time to perform multistep computations. However, this time interval can be significantly extended if the cerebellum is provided with the necessary mismatch and informational signals for learning. For example, the cerebral cortex (other higher structures) can send an informational signal to the cerebellum prior to and during the delivery of an external unconditioned stimulus in a trace conditioning paradigm. Because it is a lower OCS than the cerebral cortex, the cerebellum does not recognize the existence of the cerebral cortex, and it simply performs its task with the benefit of a richer informational context. The cerebellum is capable of learning motor tasks of greater complexity in this situation because it receives the necessary types of signals to perform the function of motor coordination.

The cerebral cortex also benefits from the cerebellum. It receives from the cerebellum the most precise information about the current state of the controlled object with regard to its trajectory of movement. In addition, because the cerebellum receives model and real afferent flow from other OCSs, the cerebral cortex is also provided with information about any mismatch between these two signals, i.e., provided with *novelty* in object behavior. The circuitry of the cerebellum can be considered to be an optimal filter that is also capable of computing mismatch signals that help to distinguish any novel stimuli that do not coincide with expected afferent flow. This may play a significant role in situations in which it is necessary to perform complex pattern recognition, i.e., when cognitive motor learning takes place. Therefore, the cerebral cortex effectively uses the cerebellum to determine in real time the state of an object whose complexity is enormous. Here we can draw one very important conclusion. The cerebral cortex also has to provide the cerebellum with model, expected, and real afferent information.

It is worth mentioning that the cerebellum is also connected with the brain stem. Some of the brain stem OCSs are functionally similar to spinal OCSs that control automatic behaviors, while others are, for instance, initiating control systems. Thus, the cerebellum is involved in the control of an object that is described by a complex hierarchy of parameters.

Obviously, the cerebellum has to have an enormous memory capacity to store its acquired knowledge. Therefore, it has to have numerous stable states and numerous attractors. The cerebellar circuitry, however, is not recurrent. How can it gain the ability to possess such a large memory? The answer is rather obvious: during each motor program, the cerebellum receives specific initiating signals that determine the task of the cerebellar circuitry, the state of the network.

The idea that the cerebellum can remember information is not new. According to Marr's original theory (Marr 1969), a Purkinje cell is a learning device switched on by its climbing fiber and capable of producing a signal that evokes "elementary movement." As a result of the increased efficacy of parallel fiber synapses, in those cases when their activity coincides with that of climbing fiber inputs, a Purkinje cell learns to respond to a signal coming via parallel fibers. Having "learned" the context during which "elementary movement" must be carried out, the Purkinje cell acquires the ability to respond subsequently to the corresponding mossy fiber signal in the absence of climbing fiber activity.

There is a major difference between this schema and the one described above. Marr and others assumed the existence of a single direction of change in synaptic weights (or more generally, the existence of a deterministic rule) under the influence of the climbing fiber signal. According to this view, the process of learning is considered to be similar to a conditioning paradigm in that it establish an association between two signals. Moreover, Marr's scheme does not explain why, after a lesion of the inferior olive, there are the same disturbances as after the lesion of the whole cerebellum. Furthermore, the nature of the "teaching signal" is unknown. For such a signal to exist, the system would have to "know" what to do in a new situation. Thus, the most complex task of deriving an optimal decision would be replaced by the qualitatively simpler task of "knowledge transmission." Last, Marr's scheme has no room for the operative function of coordinated movement correction.

In recent works, researchers have started to consider the possibility of bidirectional changes in synaptic weights between parallel fibers and the Purkinje cells during cerebellar learning initiated by climbing fibers. This idea is presently being broadly explored by numerous scientists (see for example, Kawato and Gomi 1992). However, nobody has explored the role of the model information received by the cerebellum from lower OCSs (described above) in cerebellar learning, in the process of cognition, etc.

11

The Skeletomotor Cortico-Basal Ganglia-Thalamocortical Circuit

11.1 An Anatomical Survey of Cortico-Basal Ganglia-Thalamocortical Circuits

According to earlier data, the basal ganglia receives diverse inputs from the entire cerebral cortex and "funneled" these influences via the ventral thalamus to the motor cortex (Allen and Tsukahara 1974; Evarts and Thach 1969; Kemp and Powell 1971). Subsequent studies led to a view that there is segregation of influences from the sensorimotor and association cortices through the basal ganglia-thalamocortical pathway (DeLong and Georgopoulos 1981; DeLong et al 1983). Two distinct loops have been distinguished: (1) a "motor" loop passing largely through the putamen that receives inputs from the sensorimotor cortex, and whose influences are transmitted to certain premotor areas, and (2) an "association" (or "complex") loop passing through the caudate nucleus, that receives input from the association areas, and whose influences are returned to portions of the prefrontal cortex. In recent physiological and anatomical studies, the concept of segregated basal ganglia-thalamocortical pathways has been further developed. The general principle that basal ganglia influences are transmitted only to restricted portions of the frontal lobe has been further reinforced (despite the fact that the striatum receives projections from nearly the entire neocortex).

Five basal ganglia-thalamocortical circuits have been distinguished: the "motor" (or "skeletomotor"), the "oculomotor," the "dorsolateral prefrontal," the "lateral orbitofrontal," and the "anterior cingulate" (Alexander et al 1986). According to the existing views, each basal ganglia-thalamocortical circuit receives its multiple corticostriate projections only from functionally related cortical areas. Moreover, each circuit is formed by partially overlapping corticostriate inputs, which are progressively integrated

Biological Neural Networks
Konstantin V. Baev
© 1998 Birkhäuser Boston

in their passage through the pallidum and substantia nigra (pars reticulata) to the thalamus, and from there to a definite cortical area. Usually the target area is one of those that sends projections to the basal ganglia. That is why the hypothesis appeared that the characteristic feature of all basal ganglia-thalamocortical circuits is the combination of "open" and "closed" loops.

Detailed information about the morphology of these circuits may be found in recent publications (Alexander et al 1989; Alexander and Crutcher 1990; Hoover and Strick 1993). According to current views, similar neuronal operations are performed at comparable stages of each of the five mentioned loops; the apparent uniformity of synaptic organization at corresponding levels of these loops and the parallel nature of these circuits are indirect proof of this opinion.

For the purposes of this section, it will be enough to briefly mention some details about the organization of the "motor" circuit (Figure 56a). The "skeletomotor" circuit is largely projected on the putamen, which receives projections from the motor and somatosensory circuits. In addition to these

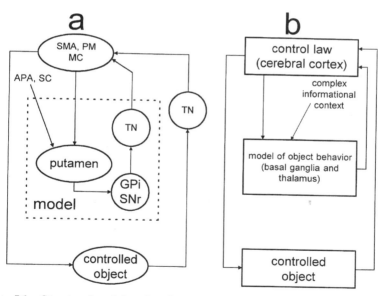

Figure 56—Structural and functional organization of the "skeletomotor" cortico-basal ganglia-thalamocortical circuit. a, "closed" and "open" cortico-basal ganglia-thalamocortical loops. Abbreviations are as follows: APA, arcuate premotor area; GPi, internal segment of globus pallidus; MC, motor cortex; PM, premotor cortex; SC, somatosensory cortex; SMA, supplementary motor area; SNr, substantia nigra pars reticulata; TN, thalamic nuclei. b, cortico-basal ganglia-thalamocortical circuit is a control system that has a model of the controlled object.

projections, the putamen also receives projections from area 5, from lateral area 6 including the arcuate premotor area, and from the supplementary motor area. While the most prominent projections of each of these cortical areas go to the putamen, there is slight encroachment of each projection upon neighboring regions of the caudate nucleus. Additional corticostriate inputs to the "motor" circuit from other functionally related regions—precentral, ventral cingulate premotor area, the supplementary somatosensory area, and certain parts of the superior and inferior parietal lobes, are still in question. The putamen sends topographically organized projections to discrete regions of the globus pallidus (e.g., the ventrolateral two-thirds of both the internal and the external segments) and to caudolateral portions of the pars reticulata of the substantia nigra. The abovementioned internal pallidal regions and substantia nigra send topographic projections to specific thalamic nuclei, including the nucleus ventralis lateralis pars oralis (VLo), to the lateral part of the nucleus ventralis anterior pars parvocellularis (VApc), to the lateral part of the nucleus ventralis anterior pars magnocellularis (VAmc), and to the centromedian nucleus (CM). The motor circuit is closed by means of thalamocortical projections from the VLo and the lateral VAmc to the supplementary motor area (SMA), from the lateral VApc (from VLo as well) to the premotor area (PM), and from the VLo and CM to the motor cortex (MC).

Thus, the general rule of connections between the cortex and the basal ganglia may be formulated in the following way: each part of the basal ganglia that constitutes a specific basal ganglia-thalamocortical loop receives inputs from much bigger regions of the cortex than those to which it projects its signals.

11.2 What is Modeled by the Skeletomotor Basal Ganglia-Thalamocortical Circuit?

Closed neuronal networks are the substrate for the model. Therefore, we may redraw Figure 56a in the other way (Figure 56b). It is necessary to note that the functional subdivision shown in Figure 56b should not be completely identified with an anatomical subdivision. This identification may be made only to a certain extent (see Chapter 2). For our purposes, however, we may do this as a first approximation and consider the basal ganglia as a system that models the controlled object.

It is not difficult to determine the controlled object of the controlling system of the "skeletomotor" cortico basal ganglia-thalamocortical circuit. It is the body of the animal and the environment. This model describes the behavior of the body and the environment during animal movements.

To predict the behavior of the object, the model uses the language of afferent signals that enter the system. For the spinal OCS, it was the language of peripheral afferents. At the level of the basal ganglia-thalamocortical circuit, the situation is different. Lower OCSs are responsible for different motor automatisms, and the problem of basic movement coordination is solved at those lower levels. Various motor control levels, such as the initiating systems of the brain stem, the cerebellum, and even the cortical level, are controlled objects for the cortico-basal ganglia-thalamocortical loop. The latter is a hierarchical control system. In addition, the cortical area to which the basal ganglia project their signals receives inputs from other cortical areas, which means that a great variety of cortical detectors (including very complex ones) sends signals to this area. These detectors determine which values have relevance to parameters describing the state of the object or its parts for any given moment in time. The model predicts the behavior of these detectors as well.

Figure 57 illustrates how the higher level of the skeletomotor cortico-basal ganglia-thalamocortical loop is functionally connected with the lower levels and cortical detectors. It is clear that there is no difference in principle between the control of lower automatisms and the control of cortical detectors. This means that the control system treats cortical detectors as subordi-

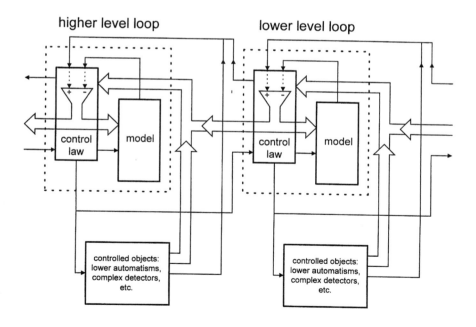

Figure 57—Hierarchy within the "skeletomotor" loop. See text for explanations.

nated control systems. Thus, the skeletomotor loop sets parameters that tune the detectors by determining definite features in a specific afferent flow.

The skeletomotor cortico-basal ganglia-thalamocortical loop is a hierarchical system. Several subloops, some subordinated to others, can be distinguished within it (Figure 57). The higher the subloop in the hierarchy, the more abstract parameters are processed by this loop. Clearly, each subloop has to receive corresponding afferent inputs to function properly. This is why all types of afferent information have to be processed by different cortical regions before afferent signals arrive in the corresponding subloop. For instance, the information from the skeletomotor subloop, which supplies the motor cortex, is not directly intermingled with information from the cerebellum, which arrives in the sensorimotor cortex. Cerebellar projections go to the cortex, which is located between motor and sensorimotor cortices.

The model for which the basal ganglia-thalamocortical circuit is the substrate is used in two ways. First, as mentioned above, it is used during execution of a movement, to predict the transition to a new state. Second, it is used during the planning of a movement, which requires a full-scale model because of the lack of afferent information about future states. It is hard to imagine this process without using a model. Moreover, a cause-effect model is not constrained by real time, and may function at rates faster than real time, e.g., rates that are necessary for fast multistep planning. It is obvious that using the model without the constraint of real time requires efferent and afferent channels to be cut off until a correct decision is reached. It is not difficult to imagine how deafferentation or deefferentation can be done at this level; different inhibitory mechanisms can be used.

The reasoning made above helps us to better understand the circuitry in the basal ganglia, and points the way to future investigations. Several simple suggestions may be made about basal ganglia circuitry. A few examples will be discussed below. As previously shown, the system at the higher level has to "jump" from one state to another while performing controlling functions. The optimal (easiest) way to create such a system is to build it on the basis of pacemaker neurons or neurons possessing bistable properties. It is easy to create a system that is able to switch from one stable state to another while simultaneously counting time intervals, by using simple trigger elements, i.e., neurons having the above-described properties. For instance, it is well known that such neurons are frequently part of the central pattern generator circuitries in many animal species. They model the period of rhythm generation during locomotion, breathing, etc.

Such neurons should be included in the basal ganglia circuitry (see Figure 58a). Otherwise, circuits including inhibition of inhibitory neurons will not work. Such processes take place in the striatum, the GPe (the external segment of the globus pallidus), the GPi (the internal segment of the globus pallidus), and the pars reticulata of the substantia nigra. Moreover, there could be one additional interesting mechanism: more complex circuit trig-

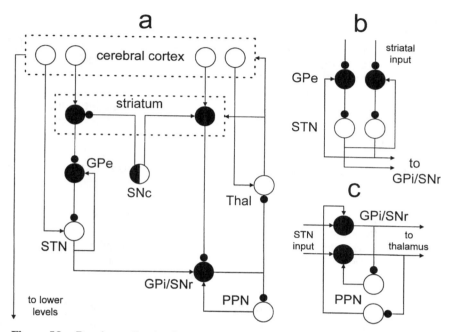

Figure 58—Basal ganglia circuitry possesses simple and complex trigger elements. a, accepted view of the basal ganglia-thalamocortical circuitry; b, c, examples of possible connections among several nuclei, which show that complex triggers may be included in the basal ganglia-thalamocortical circuitry. Excitatory and inhibitory synapses are shown by arrows and black circles, respectively. Abbreviations: GPe, external segment of globus pallidus; STN, subthalamic nucleus; SNc, substantia nigra pars compacta; GPi, internal segment of globus pallidus; SNr, substantia nigra pars reticulata; PPN, pedunculopontine nucleus; Thal, thalamus.

gers. Suppose that there are also circuitries such as those shown in Figures 58b, c. What does this mean? It means that such circuits may be in only one of two possible states. A system having a large number of such triggers, each of which may be in only one of two stable states, can encode the whole variety of possible states of the object. As can be seen from Figure 58a, such triggers may exist at different levels of the basal ganglia. Therefore, a hierarchy of states may be encoded such that higher level triggers change less frequently than lower level triggers.

11.3 An Error Distribution System

It was implied above that the control system uses the model while performing its function. As was already mentioned in Chapter 2, such systems have to have an error distribution system. It is easy to identify it in the skeleto-motor cortico-basal ganglia-thalamocortical circuit. The error distribution system includes the dopaminergic neurons of the substantia nigra, and the adjacent mesocorticolimbic group. The complex organization of the cell body subgroups—one located in the substantia nigra pars compacta and the other in the ventral tegmental area—is no longer defined in terms of striatal or mesocorticolimbic projections. These subgroups are intermingled, and some mesocorticolimbic projections have their origin in the substantia nigra, and vice versa (Le Moal and Simon 1991).

This view is consistent with numerous experimental data. Dopaminergic neurons of the pars compacta and mesencephalic tegmentum react to any behavioral or environmental change, i.e., to stimuli that are not predicted by the model. If the stimulus can be predicted, the situation is different. For instance, in conditioning paradigms, dopaminergic neurons respond to an unconditioned stimulus in the beginning of learning trials. Later, as the animal learns the task, the cells respond to a conditioned stimulus that cannot be predicted, and do not respond to an unconditioned stimulus (Ljungberg et al 1992). However, they will fire if the unconditioned stimulus is not presented at its previously predicted time interval—for instance, if the stimulus appears earlier than expected. Any previously described patterns of dopaminergic neuron firing could be misleading in terms of stimuli that activate this neuronal system, and some authors have suggested that signaling of dopaminergic neurons predicts future reinforcement (Houk et al 1995). Although the idea of prediction is still the major feature of their model, the authors ascribe this function to a circuit that includes dopaminergic neurons. As we saw above, it is more appropriate to consider the whole basal ganglia circuit as a substrate for predictions. In this case, a previous firing of dopaminergic neurons in response to an unpredicted conditioned stimulus can play a role in tuning the basal ganglia circuitry for correct future prediction of the unconditioned stimulus.

There are three possible locations where mismatch signals between model and real flows may be calculated. These mismatch signals go to the dopaminergic neurons mentioned above. The first location is in the cortex, from which mismatch signals go to those putamenal and caudate neurons that project to dopaminergic neurons. The second is in the putamen and the nucleus caudatus. The third is in both places. But it is necessary to note that there are no differences in principle among these possibilities. In all cases, an error signal resulting from mismatch between model and real flows will reliably reach dopaminergic neurons.

In addition to the above-described mismatch signals, dopaminergic neurons also receive less numerous initiating inputs from other sources, such as the entopeduncular nucleus, the dorsal nucleus of raphe, the central nucleus of the amygdala, and the bed nucleus of the stria terminalis. There are also some indications that direct cortical inputs to the substantia nigra may be present (see Brodal 1981). The existence of multiple inputs to the error distribution system does not contradict the theory. We have already seen that the cerebellar error distribution system receives inputs from the cortex that can create new minimization criteria.

These data permit one to draw the following schema of error distribution between different skeletomotor subloops (Figure 59). A mismatch signal computed at the level of one subloop goes through dopaminergic neurons to the model of the same level and to the controlling network (corresponding cerebral cortex) of a higher-level subloop.

One can see that the principle of learning described in Chapter 2 may be applied to the basal ganglia. Learning in the basal ganglia starts when initiating signals become larger than minimal, and stops when they are minimized. It is possible to suggest that very complex strategies may be used at this level to minimize initiating signals. The strategy of random search may be used in early ontogenesis, when there is no knowledge of object behavior. Later, more complex and more advanced learning strategies may be used. These strategies can be based on some specific mechanism, for instance, memorization of informational context (see Chapter 12). The complexity of cortical and basal ganglia networks makes such a suggestion reasonable. Moreover, in this system, the error signal possesses a sign: dopaminergic influence excites the

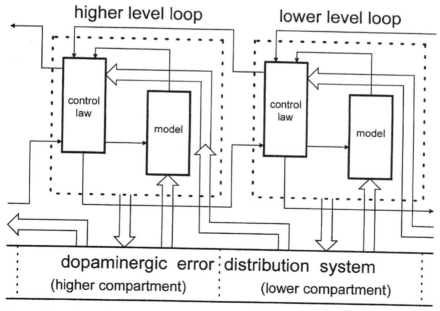

Figure 59—Dopaminergic neurons of the substantia nigra pars compacta and of the ventral tegmental area are included in the error distribution system of cortico-basal ganglia-thalamocortical loops.

direct pathway from the putamen to the globus pallidus, and inhibits the *indirect* pathway. Clearly, the sign accelerates significantly the process of learning, and rapid tuning of the model on the object becomes possible. We have already mentioned that a hierarchy of parameters exists at this level, and it is possible to make a simple analogy. Imagine that the predictive mechanism is built on the basis of a counter that encodes object states. In this case, if this predictive mechanism is working well, it will be necessary to adjust only the lower digits by changing the thresholds of their switches.

11.4 Clinical Applications of the Theory

Let us analyze what types of dysfunction will appear if there is pathology at different levels of such a system as the basal ganglia—a learning system.

The destruction of the error distribution system

This results in the partial or total loss of error signals. In the first case, the system loses the ability to slide down the gradient of the "error" signal to the optimal state, which corresponds to the minimum of "error" signal (Figure 60). As was mentioned earlier, this corresponds to the minimum of "potential energy." So the system will be in one of the local minima which is more or less far away from the global minimum. If the system is in one of the local minima, it will more or less precisely predict the state of the controlled object (more correctly, the probability of the system to be in a given state). If such a prediction is close enough to the real state, the controlling system may function quite well. The system cannot function when the prediction is absolutely wrong or even unreal, i.e., when the error distribution system is almost or completely destroyed.

Destruction at the level of informational context

If there is a diffuse nonvoluminous destruction of these channels, then the function of the system will be preserved, and only the resolution will become lower. In terms of movements, this means that the precision of movements will be reduced. Voluminous diffuse destruction may lead to the loss of function, because a large decrease in precision does not permit the system to function properly. Total deafferentation is the final stage of this process. In the case of the destruction of one specific informational input (visual, acoustic, from other parts of the body, etc.), the system loses the ability to predict the behavior of the object that is correlated with this type of informational context. For instance, if the part of the model that describes the behavior of a particular body part does not receive informational context from other body parts, any coordination between the behavior of these body

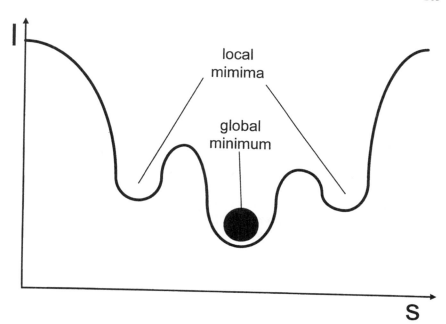

Figure 60—Simplest analogy showing that the most stable state of the system corresponds to a minimum of activity of an initiating signal. I, intensity of initiating signal; S, state of the system.

parts cannot be achieved. All other types of damage to informational channels will lead to situations which fall between the two types of damages (nonvoluminous and voluminous) discussed above. But in all these cases the model will be relearned according to the described schema of learning, i.e., the model will be returned in terms of the new situation when the system lacks a part of the informational context that it needs for the calculation of necessary minimizing function.

Damage to the modeling network

Obviously, this will lead to some loss of the computational abilities of the system, depending on the amount of damage.

   Thus, the three situations described above may sometimes lead to similar symptoms. But there are differences in the principles that govern them. In the first case, the system cannot find the global minimum, because the error distribution system does not function properly. In the second case, the system cannot find the correct output, because the informational context used for the calculation of the controlling function is incomplete. In the third case, the correct output cannot be found, because of the loss of computational abilities.

A malfunction of a neural network calculating mismatch signals

Several cases are possible:

1. A new function for minimizing an initiating signal can be calculated by the model. Therefore, there will be a final stage of the learning process in which the model network starts to calculate this function. In this case, the result of the malfunction (the expression of pathological symptoms) depends on how far this calculated function is from the correct one.
2. There is no function that minimizes the mismatch signal. For instance, the comparing mechanism generates an output signal without taking into account input signals. This case will lead to a persistent search by the system in its state space and, therefore, to the destruction of the model.

Pathologic afferent flow produced by any other subsystem of the brain begins to arrive in the normally functioning model

This will lead to a learning process that will stop when the model starts to properly describe this pathologic afferent flow and its correlations (associations) with other components of the normal afferent flow. We may call this process the "mastering" of pathology. In such a case, any associative afferent flow may provoke pathological behavior.

Only a general schema of possible dysfunction in learning systems has been described above. Obviously, analysis of all of these very complex situations would require further discussion.

11.4.1 Parkinson's Disease

This disease has been chosen as an example for several reasons. Parkinson's disease (PD) is one of the most studied neurodegenerative disorders. Clinical studies and experimental data from animal models of parkinsonism have convincingly shown that the death of dopaminergic neurons of the substantia nigra leads to Parkinson's disease. Akinesia, muscular rigidity, and tremor are the main symptoms of this disease (see Bergman et al 1990; Greene et al 1992; Montgomery et al 1991).

Thus, given the proposed theoretical approach, it is possible to conclude that Parkinson's disease is the consequence of disorders of learning processes in the basal ganglia, a disorder of the error distribution system. As a result of this, the model incorrectly predicts the state of the object in these patients (Figure 61).

Let us analyze how incorrect prediction leads to symptoms in Parkinson's disease. It is clear that the symptoms depend on which parameters of the object are predicted incorrectly (which digits of a "counter," lower or higher, are set improperly). When the state of antagonistic muscles is predicted incorrectly, there will be rigidity or tremor. In the latter case, this looks like overregulation, in which the model consistently misses the equilibrium point. More complex explanations should be used for bradykinesia. In this case, the prediction of the model differs significantly from the real state of the object. In addition, the model may predict several states with equal probabilities. Therefore, it takes much more time for the system to choose from among these states, to decide which of them is more probable.

Let us consider one of the simplest analogies—the interaction of antagonistic reflexes at the spinal level. It is well known that when receptive fields for antagonistic reflexes are stimulated simultaneously, it is impossible to predict which of these reflexes will be evoked. The latent period of the evoked reflex is usually much longer than in the normal situation. Moreover, sometimes none of these reflexes appears at all. Severe akinesia is the result of an absolutely unreal prediction of the model.

This proposed theory helps us to better understand the schema of cellular and molecular mechanisms underlying Parkinson's disease. From the theo-

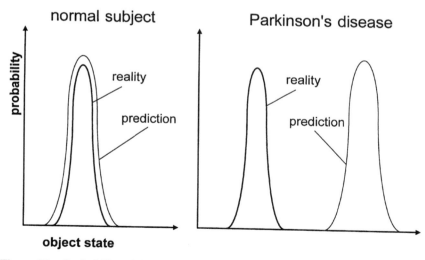

Figure 61—Probability of the controlled object being in a particular state. When the model functions correctly, the prediction coincides with the probability distribution generated by real afferent flow. Predicted and real probability distributions do not coincide when the model functions incorrectly, as in the case of Parkinson's disease.

retical point of view, there are three possible reasons for the death of dopaminergic neurons of the substantia nigra:

1. Direct destruction of dopaminergic neurons as occurs during MPTP intoxication.
2. Secondary damage to these neurons. This may happen when the molecular (intra- and extracellular) mechanisms of learning (molecular automatisms controlling learning) initiated by dopamine are not working effectively. In this case, there may be pronounced parkinsonian symptoms without significant loss of dopaminergic neurons. The activity of dopaminergic neurons will be much higher than normal in this situation, because the intensity of the mismatch signals will be increased. This overloading of dopaminergic neurons may result in their death. But this death would be a secondary result.
3. There may also be secondary death of dopaminergic cells if the neural network that calculates mismatch signals functions incorrectly, and the initiating signal is present even in those cases when there is no error in model functioning (see previous section). Such initiating signals will activate dopaminergic neurons to no purpose.

Generally speaking, in the first case there will also be secondary damage of dopaminergic neurons, because any incorrectness in the model leads to an increase in the intensity of mismatch signals. From this point of view, it is interesting to conceptualize the mechanism of levodopa therapy. In the context of the altered learning process described above, levodopa therapy functions like spurring a horse that cannot run. Acceleration of the process of deterioration of the system is the price the patient pays for temporary motor improvement. Drugs whose mechanism of action is based on MAO or COMT inhibition, or that improve molecular mechanisms of learning, seem much more promising from this point of view.

Functional neurosurgical procedures

The most puzzling problems unearthed by observations generated by functional neurosurgery for Parkinson's disease still remain unexplained. For example, why do partial lesions of particular structures within the neural network of the basal ganglia-thalamocortical loop, the globus pallidus pars interna, and some thalamic nuclei, improve symptoms (Andy et al 1963; Laitinen 1985; Laitinen et al 1992; Spiegel 1966; Tasker 1990)? This observation contradicts common sense. How is it possible to improve the network function by destroying part of it? Moreover, chronic stimulation of some of the same structures produces the identical effect (Siegfried and Lipitz 1995; Benabid et al 1991). This is a fundamental paradox that until now has not been reasonably explained by reflex theory or by the balance of excitation and inhibition that is postulated to occur in the basal ganglia. In the case of stimulation, two possible mechanisms exist: (1) stimulation produces a functional block in regions immediately adjacent to the electrode tip and (2) the spreading of influences to other brain regions occurs both *via* fibers passing through the stimulated region and *via* axons of neurons excited during stimulation. However, these latter influences do not provide the system with meaningful information for signal processing. Thus, the second mechanism simply introduces noise into the network. Therefore, an obvious question appears: why does noise added to such a system improve its function? Finally, it is well known that such a parkinsonian symptom as tremor can be effectively removed by placing a lesion in the thalamic nucleus, which conveys information from the cerebellum to the cortex. This means that the real afferent flow is changed by this procedure.

From the point of view of the theory developed above, the explanation of these medical facts could be as follows: when the network that generates model afferent flow is partially destroyed, the probability distribution of possible object states becomes lower and wider (Figure 62). Obviously, predicted states can partly overlap with possible real states of the object after this procedure. As a result, the system no longer finds an error in its prediction, and it does not try to correct the object position in its state space. The situation is the same in the case of placing a lesion in the network processing afferent flow (Figure 63). The model flow is not changed. What is

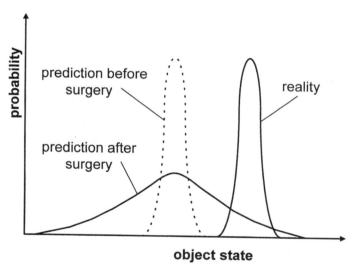

Figure 62—Effect of functional neurosurgical procedure in the case of Parkinson's disease. Partial lesion is placed in the pallidum or pallidal projections to the thalamus.

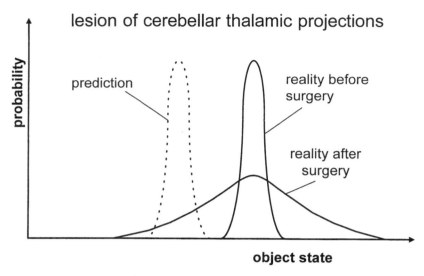

Figure 63—Effect of partial lesion of cerebellar projections to the thalamus.

changed is the probability distribution predicted by the real afferent flow; the two distributions generated by the model and real flows start to overlap. This leads to alleviation of parkinsonian symptoms, for instance, tremor.

An explanation for the effects of chronic stimulation becomes clear, from what was said above. A functional block works in the same way as partial destruction of the structure (Figure 64). Therefore, the mechanism of chronic stimulation is, in part, similar to making a lesion. Concerning noise—another possible factor that can work in the case of chronic stimulation—it is well known that adding noise helps the system to slide down the error slope to its global minimum, much as shaking an uneven sloping surface helps a ball slide down to the lowest point on that surface. While the results of chronic stimulation are similar to the clinical results of making a lesion, they do not result in immediate neural tissue destruction. Furthermore, one can stop stimulating tissue by turning off the source of current, thus producing a reversible type of functional lesion.

One critical aspect of this explanation should be kept in mind. Placement of a lesion or chronic stimulation within an OCS network improves symptoms, but does it make the controlling system function normally? The answer is no! After the lesion has been made, the system is effectively tricked and does not find any errors in its prediction. However, normal function of the controlling system is not restored. Therefore, functional neurosurgical procedures in PD should be considered merely as palliative, symptomatic interventions, and not as curative measures, because they do not stop the fundamental degenerative process. Restoration of the structural integrity of

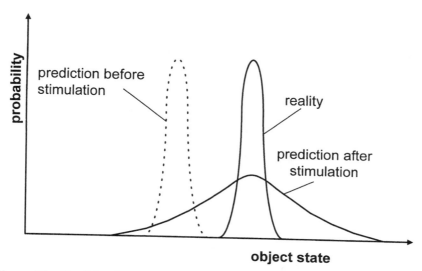

Figure 64—Possible change of predicted probability distribution under chronic stimulation of some basal ganglia nuclei (GPi, STN).

the system by rewiring lost connections, or at least prevention of further dopaminergic cell loss, would be the most effective form of treatment for PD.

Therefore, functional neurosurgical procedures must be considered to be methods of treatment based on the holographic properties of biological neural networks, which means that the system can function after partial destruction (as in holography, in which a complete three-dimensional image of an object can still be reproduced after partial destruction of the photographic plate, but with a lesser resolution).

One serious danger that must be avoided during such functional procedures as pallidotomy is damage to the prefrontal loops, which also go through the pallidum. This can be a cause of dementia, because prefrontal cortico-basal ganglia-thalamocortical loops are responsible for higher intellectual functions (see Chapter 13).

When degenerative loss of nigrostriatal dopaminergic cells was discovered to be the etiological basis of PD, the possibility of replacing them by transplantation of dopamine-producing cells—whether of neural, paraneural, or transfected cell origin (see Marciano and Greene 1992)—was embraced. It was suggested that these cells be implanted directly into the striatum in order to tonically increase local intrastriatal dopamine concentrations. From the theoretical perspective outlined in this paper, such treatment approaches appear naive and are ultimately destined to fail. The transplanted cells will not reproduce the necessary temporal dynamics of local dopamine release because nigrostriatal dopaminergic cells respond phasically only to novelty and to motivational cues.

However, several preclinical and clinical reports have demonstrated that transplantation of fetal dopaminergic cells into the striatum significantly improves parkinsonian symptoms. The question now appears: how is it possible to explain these observations from the perspective of the heuristic principles presented above? Several possible explanations may account for these observations: (1) Partial damage to the network computing the model flow can work like a functional neurosurgical procedure. The regular transplantation procedure consists of half a dozen tracks, and damage to the striatal network can be substantial. (2) A process of rewiring takes place in which new initiating circuits are produced and replace those that have been lost. In this case, the function of the initiating (error distribution) system will be improved. Given empirical evidence, this explanation seems improbable. Graft cells are implanted only in some parts of the striatal network. (3) Dopaminergic cells liberate dopamine in the intrastriatal intercellular space in the absence of rewiring. Such a result would add noise to the system. The effect of noise added to the system has already been discussed. (4) The new circuits that result from sprouting are not connection-specific. This is very similar to the third explanation, in that it is equivalent to adding noise to the system. Such nonspecific implanted circuits will act as a biological "noise generator." (5) Grafted cells initiate a process of repairing the entire system by means of function-appropriate structural reorganization similar to that which occurs in early ontogenesis. But it is highly unlikely that simple tissue implantation is capable of initiating a process of de-differentiation followed by differentiation. And finally, (6) embrionic grafted cells release trophic factors which, through humoral influences, can improve the function of any existing elements of the error distribution system.

Obviously, the above-described decrease in the resolution of the system can be effective in alleviating symptoms in other neurological diseases, such as Huntington's disease, some types of severe tremor, etc. However, it has to be well understood that such a decrease in resolution should not be very large. Let us consider the simplest example. Suppose we have a radio receiver that has high selectivity, which means that it has a high-quality resonance contour. In the normal situation, in which the resonance frequency coincides with the frequency of a radio station, the high quality of the resonance contour is a definite advantage. But it can be a disadvantage, if the receiver cannot be properly tuned to the radio station. This can happen when the receiver tunes to the station automatically with a small error that is enough to decrease the quality of reception (Figure 65). The result is familiar to everybody: the signal is distorted, and reception is unclear. Of course, it would be best to repair the receiver in this situation. If that cannot be done, however, we can still improve the quality of reception by widening the resonance curve, i.e., by decreasing the quality of the resonance contour

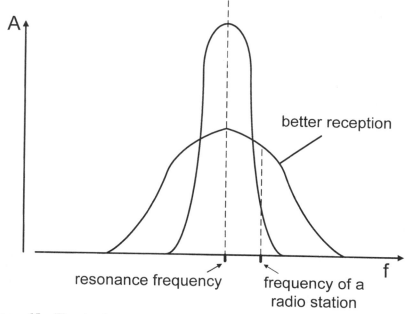

Figure 65—The simplest example demonstrating how a decrease in quality of a technical system, a radio receiver, can result in improving its function. Abscissa, frequency; ordinate, amplitude. The amplitude of a radio signal increases after the decrease in quality of the resonance contour. See text for explanations.

(Figure 65). Clearly, the latter can only be done to a certain extent. Otherwise, the receiver will lose its selectivity, and several radio stations will be heard simultaneously. This is a good example from a technical field, demonstrating that it is sometimes possible to improve some functions by means of a decrease in the quality of the system. This is what functional neurosurgical procedures are limited to achieving until other radical treatments for neurological disorders are found. A curious analogy comes to mind if we start thinking this way. Any surgical procedure leading to removal of a part of an organ, or any body part, ultimately relies on the "holographic" properties of the affected system (organ), because it results in partial loss of function that means a decrease in the resolving power of the system.

Given the suggested theoretical approach, the strategies available to functional neurosurgery become very clear. Contemporary strategies should follow a limited number of basic guidelines. The first and most crucial of these is to avoid at all costs the destruction of the substrates responsible for error distribution within a system. To destroy such a region is equivalent to the destruction of the whole controlling system. For instance, destroying only the inferior olive—a structure that functions as the substrate for error distribution in the cerebellum—leads to functional destruction of the cerebellum (see Batini et al 1985). This fact was already discussed in Chapter 10. The second guideline is to limit the magnitude of damage to substrates that (1) provide a given system with its informational context and (2) constitute the computing network. Third, in order to alleviate the symptoms of a particular disease, lesions and/or chronic stimulation should be limited to regions within a controlling system that decrease its *dysfunction* properties. Finally, it must be kept in mind that an OCS is an adaptive learning system. Because the controlling system will eventually learn to adapt to the presense of a lesion that does not *completely* destroy the output of a structure within it, it will eventually learn to accommodate itself to the attenuated information flow from residual regions within the structure that continue to function. Thus, there can eventually be a return of symptoms.

Future strategies in functional neurosurgery should be concerned first and foremost with providing mechanisms that serve to halt the process(es) underlying neurological diseases. Second, and fundamentally, these strategies should provide a means by which lost connections can be replaced and appropriately rewired in systems whose dysfunction is responsible for the genesis of symptoms of a particular disease. At the present time, modern medicine does not have the knowledge or technical capability to treat neurological disease in this manner. From the previous discourse, we can conclude that modern mechanistic transplantation approaches obviously do not possess this capacity. However, the first strategy seems more immediately attainable. Furthermore, if the latter strategy is ever realized (most probably, by influencing the cell's genome), it will surely lead to earlier diagnosis and more definitive treatments.

12

The Limbic System

The term *limbic system* usually designates the following anatomically inter-connected structures: the hypothalamus, the amygdala, the hippocampus, the septal area, and the cingulate gyrus. From a functional point of view, this classification can be justified by the fact "that many, if not all, of the effects produced by stimulation and lesions of the extrahypothalamic limbic structures can be replicated by stimulation or lesions of the hypothalamus" (Isaacson 1982, p. 2).

The limbic system is one of the most mysterious systems of the brain. It influences so many functions that presently it seems to many to be almost impossible to propose a reasonable theory that explains how this system is capable of functioning. According to numerous experimental data, the limbic system controls: activities essential for the self-preservation of the individual (including emotional ones, e.g., feeding behavior and aggression, behaviors that are often accompanied by emotions), activities essential for the preservation of the species (e.g., mating behavior, procreation, and the care of the young), visceral activities associated with both of the above and with numerous other activities of the hypothalamus, and mechanisms for memory.

The limbic system has its own hierarchy that is quite easily described. The hypothalamus and the amygdala are the lowest of the hierarchical limbic levels. The hippocampal formation constitutes the next level in the hierarchy. According to Isaacson (1982), there are two different terms—*hippocampus* and *hippocampal formation*. The first is used to designate the structures consisting of the hippocampus and the dentate area, while the term hippocampal formation is used to designate the hippocampus proper, the dentate gyrus, and the transitional cortical areas connected with both the hippocampus and the neocortical areas. For our purposes, only the second term makes sense, because we are using a functional approach. However, for simplicity, the first term will be used below to designate the hippocampal formation.

Biological Neural Networks
Konstantin V. Baev
© 1998 Birkhäuser Boston

Finally, the cingulate gyrus is the highest hierarchical level of the limbic system.

Below we shall see how each limbic hierarchical level adds new controlling abilities to the system by using the available surrounding visceral and somatic automatisms. Each of these levels has its own control specialty and semantics of afferent and efferent signals. An attempt will be made to clarify them, because understanding them will help us to comprehend the functional role of the different limbic structures. The means by which to do this will be similar to the algorithm used in previous chapters. The control system, its controlled object, the corresponding initiating and informational inputs, etc., will be described for each hierarchical level.

In order to effectively perform this analysis, one has to recognize from the very beginning that (1) there are automatisms—visceral and somatic—that are subordinated to the hypothalamus, (2) the limbic system can utilize some cortical automatisms, such as those of the skeletomotor cortico-basal ganglia-thalamocortical loop, and (3) there is a radical difference between the control regularities that were revealed for somatic automatisms, including the highest somatic levels (see previous chapters), and the control regularities at the level of the limbic system.

12.1 Associated Automatisms

12.1.1 Automatisms Subordinated to the Hypothalamus

The hypothalamus is the core of the limbic system. The best view of the hypothalamus within the framework of the proposed theory is that it is the highest-level controlling system having a nuclear structure and that it is responsible for visceral, emotional, and some somatic automatisms. In the literature, it is often designated as the "head ganglion" of those systems that are directed toward maintaining a favorable internal environment, both through the regulation of the internal organs and through the behavior of the organism as a whole.

There are numerous visceral and somatic automatisms that are regulated by the brain stem and subordinated to the hypothalamus. It is not necessary to enumerate all of them. For this discourse, it is enough to briefly mention some of them, in order to demonstrate the principle of control at the hypothalamic level. The automatisms responsible for controlling homeostasis are typical examples of visceral automatisms. Experiments have demonstrated that stimulation of some hypothalamic regions evokes changes in visceral functions such as heart rate, blood pressure, etc. Other automatisms that have a significant somatic component can also be easily evoked by electrical stimulation of the hypothalamus. They are usually designated as *elicited* behaviors. These are orientation reactions, rage reactions, licking from a water bottle, eating a food pellet, drinking water from a dish, or sexual or aggressive acts. Such elicited behavioral acts are usually described as actions that are not directed toward motivational alleviation, for instance, satisfying the need for food or water. They simply seem to be patterns of behavior that are elicited without regard to motivational satisfaction.

Given hierarchical principles and the concept of command systems discussed in Chapter 7, we can consider the hypothalamus as a region where numerous command systems for visceral and somatic control are located, although CPGs for some types of visceral control are also likely to be located in this region. Some of these command systems are higher order command systems for lower OCSs. In the case of autonomic regulation, the principle of control is no different from somatic regulation. The only difference between somatic and autonomic automatisms is the speed of the control process. The latter automatisms are often much slower than somatic automatisms. For instance, the dynamics of blood pressure regulation are slower than those of fast limb movement. Hormonal and other types of humoral regulation, such as osmotic regulation and seasonal changes, are examples of even slower processes.

It is worth mentioning that the brain itself, as a part of the body, needs a complex control system that is responsible for its own trophic regulation. For instance, recent discoveries have shown that the regulation of blood flow in the brain is intricate and complex. Modern imaging techniques such as positron emission tomography (PET) and functional magnetic resonance imaging (MRI), which have been used to reveal discretely active brain regions and their sequential changes during different mental tasks in human beings, do not directly measure neuronal activity. The changes visualized in contemporary imaging studies reflect changes in local brain hemodynamics, such as blood flow in the case of PET, and blood oxygenation in the case of functional MRI (Ungerleider 1995). Obviously, the automatisms responsible for control of the blood flow in the brain have to be very well coordinated with the changes in neuronal activity of the corresponding brain regions. The same is true for other trophic automatisms and metabolic automatisms involved in learning within the central nervous system (CNS).

12.1.2 Initiating Signals of the Hypothalamus

According to the theory, one source of initiating signals for a controlling system is the control systems that are subordinated in the network hierarchy. From the perspective of a higher command system, receiving such signals means that the lower control systems have exhausted all of their controlling resources, are incapable of performing proper control, and require a higher-level controlling influence. For instance, the system responsible for osmotic regulation can generate such a signal if it cannot support the necessary parameters of its task without having water added to the system. In other words, the animal has to find water and drink it to restore osmotic balance. In actuality, this signal can initiate a very complex series of behaviors. In the physiological literature, this type of signal is referred to as a *motivational* signal, in order to emphasize the fact that it evokes long-lasting complex behaviors that differ from the simple reaction of a lower-level control system to an initiating signal—for instance, a flexor reflex evoked by noxious stimulation of an extremity. Obviously, from a formal point of view, there is no difference in principle between the initiating signals of lower and higher levels. Regardless of the level, any initiating signal has to be minimized during control, although in the second case it can take much more time to reach a control goal. By their very nature, initiating signals exist for a much longer time at the higher level before they are minimized. Another good example of higher-level initiating signals with extended durations prior to their minimization are the internal, usually seasonal, changes that initiate very complex forms of mating behavior in animals.

An initiating signal can also be external or come from a higher system such as the neocortex or a detector having a certain level of complexity. For example, the olfactory system can send to the hypothalamus a signal about the close proximity of a potential enemy. On the one hand, the circuitry of the hypothalamus that receives this signal can be the command system, i.e., the system that initiates complex avoidance behaviors. On the other hand, the olfactory signal can be the basis for a sense of fear, a natural consequence of the property of parameter generalization in a complex hierarchical network. In this case, such a signal will initiate not only somatic automatisms, but visceral ones as well, because it is necessary to mobilize internal resources for a potential fight or for the high level of physical activity that is necessary for an effective escape reaction.

Emotion and the senses have thus far only been briefly mentioned in the context of motivational initiating signals. The senses require the higher cortical levels of the hierarchical CNS network. Concerning emotions, we have already seen that their visceral and somatic components can be executed at the level of the hypothalamus and the brain stem by means of visceral and any corresponding somatic automatisms. Therefore, although emotions in many cases require higher control levels to create their corresponding feelings and to adequately associate their expression in complex learned behaviors, an attempt will be made, in the discourse to follow, to demonstrate that their physiological basis is tightly coupled with the initiating systems of the hypothalamus and the evolution of specific types of detectors.

In order to explain this concept, let us consider some possibilities. The first is that a sensor performing an integral measure of *aversive* input signals was created in the course of evolution. What does the presence of such a sensor mean for the animal? The answer is clear. The animal will potentially be able to find the state in which this aversive input signal is minimized.

The second is, in contradistinction to the first, the creation of a sensor that performs an integral measure of *rewarding* stimuli such as tasty food, some types of olfactory stimuli, comfortable surrounding ambient temperatures, etc. The animal will thus perform different behaviors to maximize the activation of this sensor. The state in which this signal is maximal can be designated as a pleasant state. Stimulation of the anatomical region associated with activation of this sensor might result in pleasurable reactions.

Such detectors are no different in principle from those complex detectors that are responsible for the initiation of mating behavior, as was already mentioned above. Their activation can be considered to be a result of hormonal influences on specific structures within the neural network that produce the necessary initiating signals. Numerous other specific detectors also have to be involved in the control of mating behavior.

There is no need to continue to belabor the point. It is obvious that to gain new behavioral abilities, this hierarchical network level had to create new and different types of detectors. As demonstrated in Chapter 7, there has to be a match between the type of control and its corresponding detectors at each hierarchical level. In this connection, it is necessary to emphasize once more that an abstraction of parameters encoded at the appropriate hierarchical level was a prerequisite for such sophisticated improvements.

There is also another important aspect of this process worth mentioning. These new types of detectors make intra- and interspecies communication possible. This communication functions as a language—of gestures, specific vocalizations or sounds, and signal molecules—that reveals to another animal of the same or different species the intentions of the individual. Different gestures can express aggression, demonstrate peaceful intentions, signal danger, etc. In higher animals, such actions are designators of emotions. As mentioned above, their executive neural circuits are located in the brain stem in vertebrates.

Obviously, one can consider the numerous regulatory functions of the limbic system—activities essential for preservation of the individual and the species, visceral activities associated with both of the above, and numerous other activities of the hypothalamus including motivations and emotions—as automatisms activated by a specific set of initiating signals and their corresponding detectors.

In concluding this survey of the different types of hypothalamic initiating signals, it is necessary to mention the initiating signals of higher hierarchical levels in relation to the hypothalamus. Such initiating signals can vastly expand the behavioral capabilities of the hypothalamus (see below).

12.1.3 Cortical Automatisms Used by the Limbic System

In the previous section, the automatisms subordinated to the hypothalamus were briefly described. At the cortical level, there are also automatisms that can be used by the limbic system. The skeletomotor cortico-basal ganglia-thalamocortical loop is one of them. We have already seen that this controlling system possesses enough power to control the complex motor behavior of the animal by using lower motor automatisms. However, one aspect of control was glaringly absent—the external initiating inputs to this controlling system. The control repertoire of a system that does not have access to such initiating inputs is rather limited. Obviously, there should be a controlling system that determines what the skeletomotor controlling system has to do at each step of control. The role of the limbic system in this process will be discussed in the sections that follow—more exactly, the cingulate cortico-basal ganglia-thalamocortical loop. Furthermore, in order to facilitate the analysis, the primary focus of this discussion will be the initiation of cortically controlled motor behavior.

12.2 Specificity of Control Tasks: General Considerations

From what was discussed in previous sections, one can conclude that there are several features incorporating new principles of control, which have appeared over the course of evolution at the level of the limbic system. One of them is the prolonged duration of initiating signals. This feature led to the creation of specific memory mechanisms (see below). Another feature that is not present at even the highest levels controlling somatic automatisms is the following: in many cases, visceral automatisms have to utilize somatic automatisms to minimize initiating signals "born" within visceral controlling systems. For instance, different motor automatisms have to be activated to "remove" the feeling of hunger. An animal has to activate locomotor automatisms to find and catch prey and subsequently to ingest it. In this book, a broad understanding of the term *locomotion* will be used to designate any active movement of the animal or its body part in space. In contrast, somatic automatisms do not need visceral automatisms to minimize their own initiating signals. Although activation of somatic automatisms is usually accompanied by activation of visceral automatisms, such as when an increase in ventilation and heart rate accompanies locomotor activity, this involvement of visceral automatisms is supportive and is not directly necessary to achieve the goal of the control.

How does a system having such resources as lower brain stem automatisms and higher cortical automatisms (i.e., framed in a surrounding consisting of higher and lower automatisms) perform its control function? The surrounding automatisms mentioned above are the controlled object for the limbic system. Through them, the limbic system controls the body of the individual and the environment. A definition of the body that includes visceral systems, the brain itself, etc., as a controlled object is understandable. However, why is the environment included as a part of the controlled object? It is for the following reasons. During control at the level of the hypothalamus, food, water, prey, sexual partners, etc., can be considered part of the controlled object. For instance, social behavior requires the presence of other members of the same species to become a part of the limbic controlled object. The sexual activity of an individual has to be synchronized with that of other individuals of the same species as well as with environmental changes such as seasons. In this context, one has to remember that the environmental aspect of the controlled object may have a very low level of controllability, and it usually takes a great deal of time to minimize some initiating signal. Very often it does not happen at all, such as when an animal does not catch its prey or does not find a sexual partner.

While performing controlling functions, the system has to possess the necessary computational abilities and receive corresponding initiating and informational afferent inputs. In order to understand the function of the limbic system, one must try to understand what the limbic system has to be capable of doing.

12.2.1 The Coordination Problem

The problem of coordination exists everywhere in the brain where there are multiple automatisms that must be synchronized in space and time. As presented and described in Chapter 10, there are two ways to solve this problem. The first is to interconnect all possible pairs of controlling systems. In nature, this method is used only in cases where the number of systems that need coordination is not too large. The second solution is to create a coordination center that (1) receives inputs from all the control systems of lower levels, in addition to a rich informational context from higher-level detectors capable of responding to discrete features that are not available to lower levels, and (2) sends back coordination signals to all levels. The cerebellum is a typical example of the second solution.

Obviously, in the case of the limbic system, the coordination problem exists as well. For instance, numerous hypothalamic and brain stem autonomic and somatic automatisms need to be coordinated. Some visceral automatisms are extremely sophisticated. Therefore, the necessity within the limbic system for a structure that plays the role of a coordinator for hypothalamic automatisms—or a "limbic cerebellum"—becomes obvious. It will soon become clear that it is the hippocampal formation that functions in this role. However, unlike the cerebellum, the "limbic cerebellum" had to acquire a new ability in the course of evolution - the ability to remember a much longer previous history—because working time intervals are much longer at this level, and changes in system states are much slower. That means that the system has to be capable of memorizing a series of states that precede the arrival of a mismatch signal. In this case, the system becomes capable of generating control influences in advance, so that a mismatch signal is minimized. The mechanism of memorizing these "state" sequences may also be used for space and time orientation.

12.2.2 Long-Range Space and Time Orientation

The necessity for long-range space and time orientation is obvious. The limbic system has to be capable of solving navigational problems. Otherwise, an animal cannot perform purposeful movements in the surrounding environmental space because it will not know where and why to move, and will not be able to minimize corresponding initiating signals.

The ability to cope with space varies significantly among different animal species. Let us try to imagine the scheme of evolutionary improvement in the ability to navigate in space and to extend the time-control range, because it will facilitate a better understanding of the new principles of control that have to appear at the level that controls the somatic and autonomic automatisms mentioned above. Of course, such separate development of different controlling systems would be impossible in evolution because different automatisms had to develop in synchrony. However, for the purpose of this discussion, one can suggest that an animal already possesses all of the behavioral and autonomic automatisms mentioned above, but does not have an automatism for navigation. The starting point for an animal that possesses such capabilities is random searching using random movements in the environment. One can consider this search as an exploration of the environment.

In the first stage, an animal can randomly move in a surrounding environmental space without memorizing the trajectory of its movements, and can only determine the gradient of a limited number of signals such as smell, light, acoustic signals, signals from distant tactile receptors such as antennae, etc. If the animal possesses a knowledge—whether genetically predetermined or acquired by the method of classical conditioning—of correlations between the signals used to compute the gradient and the mismatch signals, it can potentially find a rewarding place or avoid dangerous regions in the future, by moving along or against the previously determined gradient. Obviously, a gradient can be very useful and can facilitate the exploration of quite a large territory. Such an animal will be, to a certain extent, capable of surviving because it can fit into some appropriate ecological niche. For many animal species, this is the only possible behavioral strategy.

In the second stage, the following strategy can be used. The reality experienced by the organism is much more complex, and not all environmental spaces can be described with the use of a gradient because the spatial topography is too complicated, such as in the case of a labyrinthine type of space. Therefore, a much more effective strategy consists of memorizing the sequence of states that preceded the minimization of each corresponding mismatch signal. In this situation, if one of the sequential states is reached by the animal during the next epoch of random searching, the final state can be found again by repeating the rest of the sequence. The controlling power of such a system will depend on the length and number of stored sequences, i.e., on memory capacity. If the capacity of memory is large enough, the animal can execute quite complex forms of behavior. The only prerequisite for this improvement in navigational capabilities is the development of detectors describing space.

In the third stage, the system acquires the ability to make temporal and spatial shortcuts, i.e., to optimize trajectories of movements in space. Theoretically, it can perform simple shortcuts during the second stage. For instance, if during explorations of the environment, the system repeats part of the sequence, the repeated part can later be excluded—a primitive or temporal shortcut (Figure 66). To make more complex shortcuts (Figure 67), the system has to possess a rather complex model of the surrounding environmental space, in which all previous experience is accumulated. Such a model possesses the knowledge of which space transitions are "permitted" and which are "prohibited." Therefore, the third stage needs some form of spatial representation, a model of space.

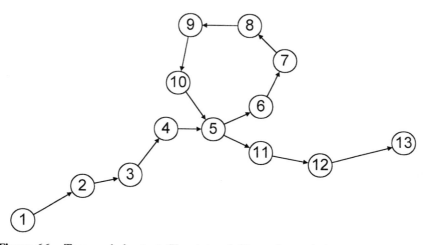

Figure 66—Temporal shortcut. The states 6–10 can be excluded during the next period of search.

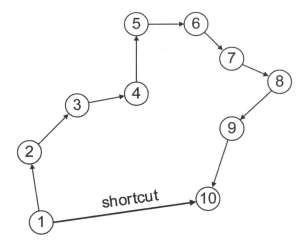

Figure 67—Example of spatial shortcut.

In the fourth stage, the ability to work within an abstract space is acquired. (see Chapter 13). This capacity is the result of further abstraction of the network's parameters and the use of features found earlier in evolution that include the ability to memorize sequences of states and to further optimize the trajectories of "motion" in such spaces.

It is necessary to emphasize once again that control processes at the level of the limbic system develop slowly. Therefore, comparatively long time intervals should be encoded and stored at this level. Different neural mechanisms can be used for this purpose. For instance, short time intervals can be counted by using an integration function of a constant signal. To compute long time intervals, the system can utilize circuit triggers or other circuitry, and cellular mechanisms such as pacemaker neurons, specific membrane properties of neurons, etc.

12.2.3 From Conditioned Reflex to Operant Learning

The description outlined in the previous section should facilitate an understanding of the nature of a qualitatively unique learning mechanism that is present at the level of the limbic system. This understanding can be accomplished by starting from the perspective of the classical conditioning schema, if one is to imagine how a conditioned reflex can be organized at this level. At first appearance, it seems very simple: a specific state reached by the object as a result of random searching in the environment should be associated with a unique and specific result—the minimization of a mismatch signal. For simplicity, one might consider a mismatch signal that emanates from hypothalamic automatisms. If the system stores only information about the final state that, for instance, coincided with punishment or an aversive state, there may be no learning advantage. It is highly possible that the system will repeat the same mistake again and again. Processes are much slower at this level, and the system may have no time for an adequate reaction to the aversive situation if it associates only the final state it achieves with the mismatch signal. In a rewarding situation, such a system is limited to performing random searching each time it desires a reward. Therefore, the classical conditioning mechanism described for lower levels has limited power, at the level of the limbic system, to solve control problems that were described for stages two through four in the previous section. However, even classical conditioning at the limbic level led to the development of a new feature. The network can compute *de novo* only an initiating signal for lower automatisms. Clearly, it is a much easier computational problem to compute *de novo* a simple organized initiating signal than to compute the whole control function realized by a lower-level automatism (Figure 68). Therefore, even at this higher level, the system also has to compute something *de novo* to create a new learned automatism according to a classical conditioning schema.

In the case of any type of conditioning, a system has to generate a prediction of future events by using information about the current state of the object. This prediction is used to compute a control signal that minimizes a particular mismatch signal. In the case of the cerebellum or other lower systems, the previous history is relatively short—tens of milliseconds—and is mainly based on primitive circuitry or cellular mechanisms such as the time constant of neuronal cell membranes and other trace processes. In the limbic system, the stored history is markedly extended—minutes, hours, and maybe even more.

The ability to store a long history can improve the learning that occurs in the schema of classical conditioning. Information from memory can be used during the interval between the event and the time of arrival of an initiating signal (Figure 68). But this is not all. It also becomes the basis for operant learning if a whole stored sequence becomes associated with an initiating signal. Obviously, if the first event of the memorized sequence is encountered again, the system can slide to the desired state by repeating the same set of consecutive states. Therefore, it is possible to formally consider this situation as *primitive operant learning*. The power of such learning, as already mentioned in the previous section, depends on the length and number of stored sequences, and the set of available automatisms that, if activated, can minimize the corresponding initiating signal. More complex types of operant learning need temporal and spatial shortcuts. It is well known that higher animals are able to perform such shortcuts. For instance, if an animal has once found a place in space where it can avoid punishment, in the future

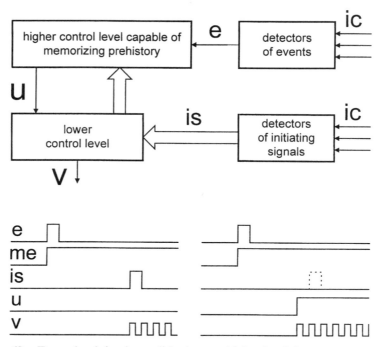

Figure 68—Example of simple conditioning at a higher level that possesses a memory. Memory information is used during the time interval between the conditioned event and the time of arrival of an initiating signal, an unconditioned event. u, v, control outputs of higher and lower control levels, respectively; e, me, event and memory of event, respectively.

it will go directly to this place regardless of its starting point in the environmental space.

It is common to distinguish the following types of instrumental conditioning methods: (1) positive reinforcement—reward, (2) negative reinforcement—punishment, (3) omission—in which the response is followed by the absence of an apetitive stimulus, and (4) escape and avoidance—in which the response is followed by the absence of an aversive stimulus (see Figure 16). The potential learning strategy described in the previous section is consistent with all these types of instrumental conditioning.

12.3 Functions of Different Limbic Structures

Now that we have examined the general considerations, it is possible to attempt to conceptualize the function of different limbic structures. We will begin with the lowest hierarchical limbic level, in order to imagine what new control abilities are added to the network system by each higher level.

12.3.1 The Hypothalamus

Initiating and informational afferent inputs to the hypothalamus come from lower automatisms and from relatively complex brain stem detectors, such as those necessary for visceral regulation, as well as for olfactory (from amygdala), visual, acoustic, vestibular, and tactile modalities.

What can the hypothalamic hierarchical level compute, having available to it only the initiating signals and informational context mentioned above, without the hippocampus, cingulate gyrus, or inputs from complex cortical detectors? Being a ganglionic structure, the hypothalamus has limited computational abilities. It can execute subordinated basic behavioral automatisms (feeding behavior, avoidance behavior, mating behavior, expression of emotions, etc.) in response to initiating signals, and can perform very limited coordination functions that are similar to intraspinal coordination functions, such as those necessary for the cross-extension of homologous limbs during walking. The abilities of the hypothalamus to cope with navigation in space are also rather limited. It is capable of localizing the source of light or sound and performing very simple types of spatiotemporal behaviors. At most, it can determine the gradient of signals from distant receptors by performing random locomotor movements for exploring the environment. This can help the animal to locate its prey, to avoid an enemy, or to find the borders of a marked territory if special signal molecules having odor, or some other signaling methods, were used for establishing territorial boundaries. However, it is difficult to imagine a circuitry in the hypothalamus that is capable of memorizing sequential spatiotemporal parameters while an animal is exploring the environment.

12.3.2 The Hippocampus

In evolution, the hypothalamus became the principal motivating source for creation of the hippocampus. The major reason for this was to acquire the advantage of access to a richer informational context from complex cortical detectors, in order to better solve the problem of coordination, and to better cope with space and time. Once created, the hippocampus began to play the role of a "visceral cerebellum."

The description of the function of the cerebellum made clear that its cortical structure resulted from the necessity of providing each cerebellar module with a rich informational context. The same is true for the hippocampus, which also has a cortical structure. In addition to the informational context available from the hypothalamus—i.e., information about the state of lower automatisms—the hippocampus receives information from various complex cortical detectors. They provide the system with detailed information about the state of the body and the position of external objects in space and time.

The use of cortical detectors implies that these detectors should be tuned on specific features present and accessible in the afferent flow. The mechanism by which this is achieved was already discussed in Chapter 7. Therefore, the fact that a vast majority of the efferent projections from the hippocampal formation go to widespread cortical regions has a logical explanation. The hippocampus performs the necessary function of tuning diverse cortical detectors by utilizing those connections. In turn, the control task for the hippocampus is determined by a variety of initiating signals from the hypothalamus, brain stem, and cerebral cortex.

The hippocampal error distribution system, which includes the septum, shares several functional similarities with the olivary system that serves the cerebellum. The septal nuclei project to the hippocampus and receive inputs from lower OCSs, for instance, from the hypothalamus, the amygdala, and the brain stem nuclei. Like the cerebellum, it has feedback from the hippocampus. Finally, like the cerebellar olive, the septum receives inputs from higher cortical levels such as the prefrontal cortex, and there are also indications that projections are present from the cingulate gyrus to the septum (Brodal 1981). Obviously, more complex minimization criteria can be created by these higher cortical levels. The olivary system projects to the cerebellar nuclei. The same is true for the septal error distribution system. The septum projects to some hypothalamic nuclei that may be considered as analogs of the cerebellar nuclei for the hippocampus. In the course of learning, septal signals have to be minimized. By utilizing the informational context available to it, the hippocampal circuit learns to compute the controlling output necessary to minimize septal mismatch signals. Any novelty in the behavior of the object will result in the activation of septal neurons.

The cerebellum loses its function in the presence of a lesion in the inferior olive. Similar effects occur in the hippocampus after septal lesions. Septally lesioned animals that are trained in avoidance tasks appear behaviorally similar to animals with hippocampal destruction. Both types of animals show an enhanced rate of acquisition in the two-way avoidance task and an impairment in the acquisition in the one-way avoidance task. It should be emphasized that the changes in learning that occur in these tasks, in animals with septal or hippocampal lesions, are always relative to the performance of normal, intact animals (see Isaakson 1982).

In light of the discourse made in section 12.2, one can hypothesize that the hippocampus is a structure that memorizes temporal sequences of various states in order to perform its control function. Such states can include visceral states as well as states encoding environmental spatiotemporal relationships. It is well known that the hippocampal circuitry possesses recurrent connections. This attribute is utilized in numerous network simulations, and it has been demonstrated that such recurrent networks are capable of memorizing sequences and of performing temporal shortcuts (Levy 1989; Levy et al 1995).

As we saw in Chapter 10, the cerebellum receives model and real afferent information flow. It is thus possible to propose that the hippocampus also receives these two types of information. If this suggestion is correct, then the hippocampus is also capable of computing mismatches between real and model afferent flows, i.e., reacting to novelty in afferent information.

The cortex is provided by the cerebellum with the knowledge of movement trajectory. The same is true for the hippocampus. The latter sends its ouput signals to lower OCSs and to the cortex, to the cingulate gyrus, and to the prefrontal cortex. From a formal point of view, this means that the cerebral cortex is provided with the knowledge of any possible "movement" trajectory determined by the hippocampus, and the cortical level can subsequently optimize this trajectory by using optimization criteria not available at the hippocampal level. The result of this optimization is the ability to perform spatial shortcuts, as mentioned in Section 12.2. The foregoing discussion facilitates the understanding of the hippocampal mechanisms utilized for recent memory (see next section).

It becomes perfectly clear, from the conceptual description above, that complex network, cellular, and molecular automatisms must be behind the ability to memorize sequencies. This is why experimentally found long-lasting changes in such networks (for example, long-term potentiation) find an easy explanation within the limits of the proposed concept.

12.3.3 The Cingulate Gyrus and its Cortico-Basal Ganglia-Thalamocortical Loop

Before starting any analysis of the function of the cingulate gyrus, one must first remember that it is a part of the neocortex associated with the basal ganglia-thalamocortical loop, a system that possesses powerful predictive abilities. This loop goes through the so-called ventral striatum.

The existence of the ventral striopallidal system was first proposed in 1975 by Heimer and Wilson. Striking parallels between the connections and histochemical features of the neostriatum and those of the nucleus accumbens and the medium-cell portion of the olfactory tubercle were identified (Heimer 1978; Heimer and Wilson 1975; Nauta 1979). The ventral striatum, the nucleus accumbens, and the olfactory tubercle receive extensive projections from so-called "limbic" structures, including the hippocampus, the amygdala, and the entorhinal (area 28) and perirhinal (area 35) cortices (Heimer and Wilson 1975; Hemphill et al 1981; Kelley and Domesick 1982;

Kelley et al 1982; Krayniak et al 1981; Nauta 1961). This is why this portion of the striatum has been referred to as the "limbic" striatum (Nauta and Domesick 1984). While initially the ventral striatum was believed to receive its cortical input exclusively from nonisocortical areas, later experiments revealed that there are also significant projections to it from the anterior cingulate gyrus (area 24) and widespread regions of the temporal neocortex, including the temporal pole and the superior and inferior temporal gyri. There are also indications that the nucleus accumbens receives projections from posterior portions of the medial orbitofrontal cortex (area 11).

The "limbic" striatum projects to the precomissural (ventral) pallidum, the rostrolateral region of the internal pallidal segment, and the rostrodorsal substantia nigra. These structures project, in turn, to posterior and medial portions of the thalamic mediodorsal nucleus, whose projections to the anterior cingulate gyrus partially close the loop. Projections to the centrum medianum nucleus of the thalamus and to the reticular formation have also been described.

A crucial feature of any cortico-basal ganglia-thalamocortical loop is the presence of mutual functional interconnections between cortical regions that send their signals to the loop (see Chapter 11). The same is true for the "limbic" loop. The cingulate gyrus has mutual functional interconnections with all cortical regions participating in the "limbic" loop. Thus, the cingulate gyrus can determine the functional control tasks of various cortical detectors, and in addition to the information it receives *directly* from them, the cingulate gyrus also receives from the loop a model of what it should *expect* to receive.

As described in Chapter 11, the cortico-basal ganglia-thalamocortical loops have dopaminergic inputs that serve to provide the network with mismatch signals. The same is true for the anterior cingulate gyrus. According to anatomical data, the anterior cingulate area receives dopaminergic inputs (Felten and Sladek 1993). Therefore, one can draw one very important conclusion: the dopaminergic input to the cingulate gyrus is the source of mismatch signals, and fast dopaminergic learning plays an important role at this control level. The cingulate gyrus was probably the first cortical level that began utilizing this type of learning in evolution.

One of the early techniques used to study neuronal interconnections, strychnine neuronography, revealed widespread interconnections of the cingulate gyrus with other brain regions. The application of strychnine to any part of the cingulate cortex can induce responses in all other structures that constitute the cingulate system (Pribram and MacLean 1953). Thus, interconnections exist among all parts of the cingulate system. Strychnine application to the posterior cingulate region evokes activity in the precuneate region and in the posterior hippocampal areas as well. The most prominant responses were those of the subicular areas, suggesting that distinct subdivisions of the hippocampus are connected with discrete regions of the cingulate gyrus. Strychnine stimulation of the anterior cingulate regions produces activity in the motor cortex and the prefrontal and orbitofrontal corticies (Dunsmore and Lennox 1950). Projections to the centrum medianum nucleus of the thalamus and the reticular formation have also been found by means of stimulation techniques (French et al 1955). There are also projections from the cingulate gyrus to the brain stem.

Several connecting regions of the premotor areas are located in the cingulate sulcus, and include the caudal cingulate motor area on the dorsal bank (CMAd), the caudal cingulate motor area on the ventral bank (CMAv), and the rostral cingulate motor area (CMAr). It has also been shown that these motor areas, as well as the supplementary motor area (also located on the medial wall of the hemisphere), project to cervical and lumbosacral segments of the spinal cord (He et al 1995). Three of these areas, the SMA, CMAd, and CMAv, resemble the primary motor cortex because of their distinct arm and leg representations. In each motor area, the size of the distal representation of a particular limb is comparable with the size of its proximal representation. It is proposed that this finding means that premotor areas located on the medial wall of the hemisphere may be involved in the control of both proximal and distal limb movements.

It has been shown that the cingulate motor areas and the supplementary motor area are connected with the primary motor and sensory cortices (Darian-Smith et al 1993). There are also interconnections among different premotor areas. Each separate region of the premotor area is connected to one or more different premotor regions. In addition, five of the six premotor areas in the frontal lobe are interconnected with the dorsolateral prefrontal cortex (Lu et al 1994). The existence of several premotor regions in the cingulate gyrus, and the complex structure of its connections with various brain regions that have different modalities, imply the existence of a complex hierarchy within the cingulate gyrus, which will not be described here.

Therefore, the cingulate gyrus has all that is necessary to perform control tasks for cortical and lower-level motor, and perhaps several other, automatisms with different functional modalities, because it possesses the necessary informational inputs and outputs (Figure 69). Portions of this informational input come from the brain stem level through the MD (mediodorsal thalamic nucleus), AD (anterodorsal thalamic nucleus), VM (ventromedial thalamic nucleus), and LD (laterodorsal thalamic nucleus). Finally, the initiating inputs available to the cingulate gyrus must be identified. It is possible to suggest that these inputs emanate from the AV, the anteroventral thalamic nucleus. This thalamic nucleus receives inputs from error distribution systems such as the septum, the mammilary bodies (a very interesting structure that is connected with both the septum and the hippocampus), and from the hippocampus itself. However, it is highly possible that some initiating signals arrive in the cingulate gyrus through other thalamic nuclei, or directly from the subiculum, because these initiating signals might be the

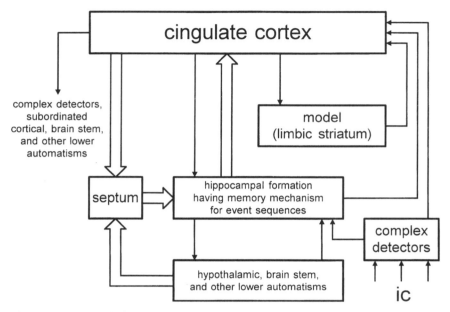

Figure 69—Interrelations among different levels of the limbic system. The principle of connections of the hippocampal formation with lower and higher levels is very similar to the principle of cerebellar connections with its lower and higher control levels. The major difference consists of the fact that the hippocampal formation possesses special memory mechanisms. See text for explanations.

result of information preprocessing in the hippocampus. The prefrontal cortex is also a powerful source of initiating inputs for the cingulate gyrus.

To continue the analogy of the hippocampus with the cerebellum, it is worth mentioning once again that the cingulate gyrus can (1) establish and initiate control tasks for the hippocampus, (2) create the necessary informational context for the hippocampus by tuning complex cortical detectors, and (3) create new minimization criteria for the hippocampus by influencing it through the septum, the analog of the inferior olive. As previously mentioned, the latter possibility is included because there are indications in the literature that the cingulate gyrus has connections with the septum (see Brodal 1981). In addition, the prefrontal cortex (see Section 12.3.2) also sends signals to the septum. Obviously, the very complex criteria computed by a fast learning subsystem utilizing dopamine-mediated learning can provide a powerful source of informational and mismatch signals that eventually results in an expansion of novelty and other unique criteria for the control system.

Let's return our attention to the problems raised in Section 12.2. The cingulate gyrus with the limbic loop and the initiating and informational inputs described above reached a new level of complexity and abstraction of its encoded parameters. For the cingulate gyrus, the hippocampus is a portion of its controlled object. Together with the hippocampus, the cingulate gyrus can make spatial shortcuts, store information about potentially permissible and impermissible space transitions, compute transitions by using gradients, etc. Such features simply demonstrate that the limbic system can effectively cope with the surrounding spatial environment. This also means that there is no longer a need for the detailed knowledge of the sequence of states memorized by the hippocampus when the cingulate gyrus with its limbic loop memorizes an optimal trajectory, a shortcut. This explains quite well the role of the hippocampus in short-term memory.

Certain experimental facts and theoretical data are worth mentioning in connection with the discourse made earlier in Chapter 12. In freely moving animals, single unit recording has revealed *place cells* in hippocampal fields CA3 and CA1 (see Burgess et al 1995). Their firing occurs only when a rat is in specific small regions of its environment referred to as *place fields*. The firing properties of place cells can be manipulated by changing an animal's environment, by changing the relative positions of pertinent environmental cues. In more complex spaces such as mazes, place cell firing depends on the direction that an animal travels, as well as on the animal's location. Spatially-correlated cell firing has also been identified in the entorhinal cortex and the dorsal presubiculum. However, their place or directional correlates tend to be more complex than those of CA3 and CA1 place neurons.

Hippocampal lesions lead to different consequences in different animal species. In human beings, bilateral damage of the hippocampus and its nearby structures, as is occasionally done surgically in epilepsy, to stop seizure produces profound retrograde and anterograde amnesia. In simpler animals such as rats, it results primarily in a deficit in spatial navigation.

Lesions made in the cingulate cortex lead to the disruption of the orderly sequencing of behaviors (see Isaacson 1982). This finding is well documented for different animal behavioral paradigms. In human beings, surgical disruption of the cingulate system, performed for the relief of pain, leads to similar impairments. In tests of cognitive function, the greatest changes are in nonverbal tasks that require temporal ordering or sequencing.

Several formal and neural network approaches seem to be successful in describing certain aspects involved in the sequencing of behavior and navigation. Tolman (1932) was the first to suggest that animals acquire an expectancy that the performance of response R1 in situation S1 will be followed by a change to situation S2. He also hypothesized that a large number of local expectancies can be combined into a *cognitive map*, a set of connected places that are systematically related to each other by a distinct set of spatial transformation rules. According to some theories, spatial navigation can be performed using either a cognitive map or a route system such as a list of Tolmanian situation-response-situation instructions (Schmajuk 1995;

Schmajuk et al 1993). Computer simulations based on these concepts have shown that a network can successfully describe the latent learning and detour behavior that occurs in rats. Moreover, what is even more amazing, is the findings of Schmajuk and Thieme (1992) who applied the Tower of Hanoi task to the network simulation, a task that is often used to test the function of the prefrontal lobes and procedural memory in the basal ganglia in man. Their simulations demonstrated that the network takes only a few trials to solve the problem in a minimal number of steps. These theoretical concepts and experimental findings are consistent with what has been hypothesized and described in this chapter. In order to solve control problems at the level of the limbic system, its control network must be capable of memorizing sequences in any form, by the use of *lists, stack-types* of memory, etc., and it must possess a knowledge of possible space transitions or "movements" in order to further optimize a trajectory of "movement" (to make a shortcut).

13

The Prefrontal Cortex

13.1 Means for Further Evolutionary Improvements

At this point in the description of this conceptual theory of higher-level brain function, one must ask a very important question: what new functional capabilities evolved with the creation of the limbic system level? The answer would be quite clear. The system acquired the ability to actively learn and perform discrete multistep control by utilizing the memorization of intermediate states and the process of control optimization.

Thus, at first sight, it appears as though the system has become a powerful and intelligent universal network computer that has been provided with an external memory. This situation is reminiscent of Turing's machine, introduced by Alan Turing. This machine exhibits many of the features commonly associated with a modern computer. This is no accident because Turing's machine provided the model for the design and development of a programmable modern computer. A Turing machine may access and alter any memory position by utilizing a sequence of elementary operations. It has no limitations on the amount of time and memory available for a computation. The presence of a feedback connection with an external memorizing apparatus having infinite memory is the feature that distinguishes Turing's machine from the simplest automatic apparatuses having memory, the so-called finite automata.

Therefore, the brain without the cerebral cortex, but with peripheral afferent feedback, could be considered as a Turing's machine. The environment, as a controlled object (the object of action from the organism), has a practically infinite set of states. By performing changes in it, the animal can recognize these states with the help of the brain, and on the basis of the obtained results, can perform further actions (for instance, build a nest). But the environment can change under the influence of different factors independent of the animal, unlike the memory in Turing's machine. Moreover,

Biological Neural Networks
Konstantin V. Baev
© 1998 Birkhäuser Boston

the realization of most complex functions with the help of recursion and minimization operators makes it necessary to operate on the basis of decisions made in previous situations, and also to examine possible variants until a satisfying variant is found. This is exactly what Turing's theoretical machines never achieved! Their memorizing apparatus does not have its own dynamics.

We have also discussed the role of the abstraction of parameters for a hierarchical network computer. It is obvious that if the level of the parameters' abstraction in such a system became sophisticated enough to perform the "movement" of abstract objects in an abstract state space, it would clearly indicate the creation of a new functional principle, and the process of elaborately detailed multistep planning would be possible, i.e., the capacity for *thought*. However, if the computational abilities of the limbic system are analyzed more carefully, one might conclude that they are rather limited. The spectrum of available initiating and informational inputs that may be utilized by the limbic system is the source of its computational and functional limitations. The concepts described in the previous section support the contention that the minimization of initiating signals born within the hypothalamus is sufficient for the principal inborn needs of the animal. The same is true for the informational context from complex detectors that are used to solve the problems of visceral control and navigation, because they are tuned on particular object features that coincide with the needs of a given control task. Finally, when the functional features required to solve the problem of environmental navigation were analyzed earlier, it was implied that the environmental cues present within a space to be explored do not change, i.e., the surrounding physical environment is static. It is necessary for an animal in a *dynamic* environment to have a knowledge of environmental dynamics in the form of an internal model. The more complete the model, the more adaptive is the behavior of the animal. It is clear that the highest-level control problems are not always solved by networks that function as reactive systems, although that is the final stage of mastering an automatism in many situations. As mentioned before, a complex control task needs recursive computations that cease once its specific criterion is satisfied. The creation of new minimization criteria requires a model on a full and elaborate scale, and significantly depends on the quality and completeness of the model.

One of the possible means of reaching a new level of control abilities, in accordance with the principle described above, is the creation of additional hierarchical levels that have available to them the necessary informational context and initiating signals. These levels also have to be able to use hippocampal circuitry in order to store the results of the intermediate computations that transpire during the process of learning. An additional hippocampal subdivision dedicated to serving the needs of this level also has to be created. Ultimately, this new level, to some degree, should be a further hierarchical extension of the anterior cingulate gyrus—a system that utilizes fast dopamine-mediated learning, an advantageous evolutionary process that resulted in the acquisition of the following features: (1) an increase in the number of intermediate computational steps that might be necessary for minimizing a visceral initiating signal, and (2) the ability to create new and more complex and sophisticated classes of initiating and mismatch signals for directing the control tasks of lower levels. These new initiating signals may be conceptualized as higher-order derivatives of fundamental visceral initiating signals, and as capable of initiating complex behaviors. However, the fundamental initiating signals themselves will continue to localize their origins in visceral control systems. For instance, a person can exhibit very complex forms of social behavior in order to satisfy needs for food, pleasure, etc. Additional features that were acquired during this expansive evolutionary process include: (1) the ability to tune cortical and other detectors on any desirable criterion, (2) the creation of memory processes that are capable of storing the most precise and complete model of an environment's most abstracted cause-effect relationships, so that the process of random searching is replaced by the process of sorting out all of the possible variants of a control task, (3) the capability of processing any environmental novelty at this level, and (4) the ability to function independently of real time so that the new level can actively be in the process of computation while lower levels are occupied with the control of ongoing current behavior.

Clearly, other cortical and subcortical automatisms are major components of the controlled object of this new level. It is not difficult to identify this control level in the anatomical structure of the brains of higher vertebrates. It is the prefrontal cortex with its cortico-basal ganglia-thalamocortical loops. This is exactly what was created by nature. It found, as usual, a "very simple" solution that reminds one of "internal tape" with one significant difference from Turing's theoretical machine—it has its own dynamics. Although this "tape" is not infinite, it has an enormous memory capacity. As we will see below, this and some other limitations forced man to begin using external memory.

13.2 Prefrontal Cortico-Basal Ganglia-Thalamocortical Loops

As mentioned in Chapter 11, the existence of a single "complex" or "association" loop, passing through the caudate nucleus and influencing certain prefrontal "association" areas, was proposed more than a decade ago (DeLong and Georgopulos 1981; DeLong et al 1983). Subsequent anatomical studies have shown that there are at least two distinct cortico-basal ganglia-thalamocortical circuits that selectively project to separate prefrontal regions: (1) the dorsolateral prefrontal and (2) the lateral orbitofrontal circuits. The corresponding prefrontal areas provide "closed loop" input to the corticostriate circuit. In primates, the open loop portions of this circuit include the posterior parietal cortex and the arcuate premotor area for the dorsolateral prefrontal loop, and the superior temporal gyrus, inferior temporal gyrus, and anterior cingulate area for the lateral orbitofrontal loop. Both loops also project to the globus pallidus and substantia nigra *pars reticula*, and from these structures to the thalamic nuclei. The dorsolateral prefrontal loop is closed *via* the parvocellular portions of the nucleus ventralis anterior and the nucleus dorsalis medialis. The lateral orbitofrontal loop closes by passing through the magnocellular portions of these nuclei. It is highly probable that new and distinct subdivisions of the association loops will arise anatomically in the future. For the purpose of this discourse, it is sufficient to consider one "association" loop that contains within it an intraloop hierarchy, as described in Chapter 11. The error distribution system for this association loop is also organized in a manner that is similar to the nonassociation cortico-basal ganglia-thalamocortical loops. It is necessary to mention that dopaminergic neurons of the mesencephalic tegmentum, which includes the substantia nigra *pars compacta*, also receive inputs from the brain stem reticular formation, the locus ceruleus, etc. These structures may conceptually be considered to function as brain stem detectors of novel stimuli. For example, noradrenergic neurons of the locus ceruleus are known to fire in response to novel stimuli (Cerbone and Sadile 1994; Vankov et al 1995).

The prefrontal cortex is widely connected with other cortical and subcortical structures and, as a rule, these connections are reciprocal. The prefrontal cortex shares connections with the parietal, occipital and temporal cortices. It also sends signals to the hypothalamus, so that inborn initiating systems that are capable of sending long-lasting initiating signals can be activated. Obviously, any voluntary minimization criterion can be created by the prefrontal cortex, and as a result the animal will feel, for instance, pleasure when this signal is minimized. Additional novelty criteria can be sent to the hippocampus through the septum. Finally, detectors of any desirable complexity can be created by tuning the appropriate cortical detectors, for instance, in the parietal, occipital, or temporal lobes. The control tasks of the cingulate gyrus and the hippocampus can also be designated by the prefrontal cortex.

Therefore, all surrounding cortical and subcortical automatisms are available to the prefrontal cortex. The most interesting feature of the prefrontal cortex is that it is capable of utilizing its basal ganglia loops and the limbic system, and its hippocampal and cingulate structures, to store the results of intermediate computations while performing random searching in its state space, which may be much more abstract than the working spaces of systems analyzed in previous sections. As noted earlier, the method of random searching by trial and error is the most powerful tool available for identifying correct decisions when the environmental situation is unfamiliar, and at the level of the limbic system this method acquired the new ability to store the results of intermediate computations. This ability is one of the most important functions of the prefrontal cortex.

Let us consider some illustrative motor behavioral scenarios in order to better understand the principles of control at this level. Suppose an animal has a goal and flawlessly performs the necessary acquired movements in a familiar environment, to achieve this goal. If some new object appears in the environment, the prefrontal cortex will immediately receive mismatch signals from one or several sources. These signals act as an interruption of the current ongoing goal-directed computation. The prefrontal cortex will immediately begin to analyze this novel stimulus, using any or all available informational context by properly tuning numerous cortical detectors and by retrieving information stored in the previous model of the familiar environment. Several possibilities exist. The first is that the object and its significance are recognized. In this case, the knowledge stored in the form of the control law and the environmental model contained within the controlling system results in (1) a continuation of the animal's current behavior, if appropriate to the new stimulus, or (2) a change in the animal's behavioral program to one that is already a part of its behavioral repertoire. Whether or not a behavioral change transpires depends on the significance of the new object with respect to the animal's initial goal. The second possibility is that the object is absolutely new and the animal has no knowledge of this new object behavior based on what it has learned of novel environmental objects encountered in the past. In this situation, the power of the entire prefrontal system and its subordinated automatisms can be used to solve the problem, and this process can take a lot of time. However, the animal's first reaction may be to activate an automatism that it has already learned, such as to run away in unknown and potentially threatening situations, and to observe the behavior of the object from a safe distance. The dopaminergic error distribution system functions in such a way that any novel stimulus can potentially evoke learning. This has a profound consequence in that the system is capable of learning by watching the behavior of environmental objects.

The regularities of information distribution across a hierarchical neural network medium such as the prefrontal cortex and its basal ganglia-thalamocortical loops facilitate an understanding of the involvement of the prefrontal cortex in the mechanism responsible for epileptic seizure activity. One must first recollect how a mismatch signal from a low hierarchical level gains access to higher levels. It will ascend higher and higher until it encounters a competent level for minimization. In this connection, the focus of seizure activity—a *seizure generator*—may be conceptualized in the following manner.

A seizure focus generates an unpredictable signal for the system, i.e., it generates noise within the system, and one can think of it as a source of pertubation to a controlling function, which is analogous, to a certain extent, to the effects on the network of introducing a source of perturbation—for instance, movement perturbation. For the control system, this perturbation is a form of novelty and is unpredictable, or has at the very least a very low level of predictability. Obviously, the control system has to exert much more effort to perform an appropriate control task to accommodate this noisy signal, and it receives mismatch signals that can spread simultaneously across several levels each time the perturbation occurs. Therefore, there will be an additional computational load on the learning systems involved in tuning both controlling and model networks, and if this load is high, sooner or later the available learning automatisms will be exhausted within a given hierarchical level. The latter process may subsequently lead to more and more pronounced spreading of the mismatch signal beyond its original limits. In this situation, the mismatch signal becomes a powerful source of an initiating signal that has to be minimized by the whole control system. Eventually, all available levels become involved in this minimization process, and seizures occur at this stage. Minimization of the initiating signal occurs later as a result of seizure generator fatigue. The state necessary for minimization of this signal, and the way to achieve that state, are memorized by the system, and each time the seizure generator becomes active again, the system will reach this state with greater ease. Thus, one can consider this process to be the end result of learning that leads to the development of a *seizure automatism*. The logical extension of this hypothesis is that dopaminergic transmission may be involved in seizure activity, a fact that has been shown experimentally (Ferraro et al 1991; al-Tajir and Starr 1991). The mechanism described above is also consistent with the finding that frequent seizures inevitably lead to dementia. Overactivity of the dopaminergic error distribution system during the seizure episode results in the deletion of information that was being accumulated by the network in the course of previous learning. Interestingly, it appears as though a neural network that incorporates fast learning mechanisms is very vulnerable to seizures, and that perhaps this is the price the brain must pay for its capacity for fast learning, which was imparted to it as a consequence of a significant adaptation developed by nature in the course of evolution.

Many neurological disorders might be considered to be pathological automatisms created at the highest brain levels. Tourette's Syndrome and obsessive-compulsive behavior are examples of such automatisms. Obsessive-compulsive behavior can be conceptualized as a situation in which an abnormal initiating signal is produced at a particular hierarchical level. This situation means that this level is not capable of leaving a corresponding region in its state space. Obviously, a theoretical analysis of possible mechanisms leading to the formation of abnormal automatisms in such a hierarchical system as cortico-basal ganglia-thalamocortical loops can help to reveal the very nature of this and other disorders. As we saw, malfunctions in such a system can occur in different places—in the controller, in the model, and in the error distribution system.

14

Conclusion

We can now make a natural and logical step toward further generalization of the automatism concept. Obviously, it is possible to talk about different automatisms in addition to the automatisms of neural networks. They can be molecular, genetic, biochemical, cellular, and even social. In all these cases, the control system is hierarchical, and each level is an OCS built of a network of interacting elements. Therefore, all of the concepts described in the previous chapters are applicable to non-neuronal automatisms. This conclusion is completely in line with the well-known fact that the same type of computation can be realized by using different types of elements—electronic devices, mechanical devices, chemical reactions, etc. Some examples of non-neuronal networks are discussed below in order to show how such a generalization is broadly applicable. Let us first go back to the process of learning in the nervous system, taking into account the presence of molecular and other non-neuronal automatisms.

14.1 The Variety of Memory Mechanisms in the Brain

Complex intracellular processes are apparently initiated in order to maintain the functioning of memory mechanisms. However, the direct encoding of functional modalities such as auditory and visual information in a macromolecular structure, as has been suggested in certain theories of memory, is considered to be highly unlikely. Within the framework of the ideas and concepts developed in previous chapters, it is sufficient for information on the current states of a given nerve cell to be stored in the macromolecular structure.

So far, we have only been discussing two types of memory-linked alteration in neural networks, consisting of (1) the transition from one stable state of network activity to another (i.e., operative or short-term memory, a

Biological Neural Networks
Konstantin V. Baev
© 1998 Birkhäuser Boston

metastable equilibrium state), and (2) plastic changes in neuronal synapses (long-term memory). This simplification is justified by the fact that structural and functional cerebral organization and the general pattern of processes occurring in the brain are given top priority. A few thoughts have to be added about the multiformity of memory mechanisms.

The most influential researchers in the neurobiology of memory attempt to provide a more or less complex classification of memorization processes, usually comprising intermediate categories aside from those of long- and short-term memory. The concept of an OCS actually suggests that optimization takes place and involves all modifiable system parameters. In particular, it is not just the functional structure of the OCS, but also its physical substrates, that must be optimum. It therefore follows that there must be as many different memory mechanisms as there are different types of system parameters. When the optimized system is, for instance, a neural network, the memory mechanisms may be based on long-term change in characteristics specific to the neuron. Such distinguishing characteristics might include excitation threshold, time constant of change in the aforementioned parameter, duration of its refractory period, membrane size and shape, synapses, synaptic distribution on its dendrites and soma, sites of neuronal axonal terminals on other neurons, parameters of intracellular processes, and so on. In addition, any characteristics of glial cells that are associated in any way with the parameters of the neuronal network may serve in some manner as a basis for the memory mechanisms in question. Memory mechanisms themselves should thus be viewed as an OCS specializing in controlling changes in the parameters enumerated above. Numerous hierarchically subordinated automatisms can be included in such an OCS, and the descent of automatisms described in Chapter 7 will take place in such a system during learning. Depending on the speed of such changes, these OCSs may have a neuronal substrate such as a neuromodulatory system, or may be based purely on humoral regulation. Lastly, taking into account that differently organized OCSs must, as a rule, be closely linked with reciprocal control influences over one another's parameters, we can begin to imagine the extent of the complexity of organization of the OCS referred to as the brain.

14.2 Non-Neuronal Network Cellular and Molecular Systems

Two examples, (1) the example of the immune system and (2) some intracellular processes, are discussed below in order to demonstrate that the concept of automatisms is applicable to them. The first example appears to be more convincing than that of intracellular processes, because the immune system has been investigated sufficiently to elucidate the scheme of its functional organization. As for intracellular processes, current knowledge about them is incomplete, and we have only a limited understanding of its organization.

14.2.1 The Immune System

The main task of immune automatisms (within the limits of the preservation of the internal integrity of the organism) is defense of the organism from intrusion by "alien" forms of life. Among the different species of the animal kingdom, this OCS is most developed in vertebrates and has the most effective subsystems (automatisms). Two aspects of the function of the immune system—the principles of clonal selection and immune memory—are discussed in this section.

In the middle of the twentieth century, scientists still could not adequately explain why the immune system is able to produce antibodies to numerous antigens—those foreign substances able to evoke an immune response. The problem was that there is an almost inconceivable variety of antigens and each type of antibody can neutralize antigens of only one type. Antigens differ in shape, size, and chemical composition. Examples of antigens are bacteria and their toxins, viruses, pathogenic fungi, parasites, artificial chemical compounds, etc. Clonal selection theory, which by now has been confirmed by numerous experiments and widely accepted, explains very well the process of antibody production (Alberts et al 1994).

According to this theory, each lymphocyte, in the process of its development, acquires the ability to react specifically with a specific antigen, even though the lymphocyte has never had any contact with the antigen. Protein receptors at the cell surface are capable of interacting specifically with a given antigen. The binding of the antigen to such a receptor stimulates increased production of antigen-specific lymphocytes. If they are B-lymphocytes, they then begin to synthesize antibodies, which are a free form of antigen-specific membrane receptors. Thus, selective cell stimulation is the basis for the antigen-specific immune response.

The selection process described above is referred to as "clonal," because according to the clonal selection theory, the immune system includes millions of different clones (cell families). Cells of one clone have similar antigen specificity because they are descendants of a single ancestor cell.

As should be expected in a system functioning according to the principle of clonal selection, even an antigen having one antigenic determinant (the portion of the antigen interacting with the antigen-binding part of an antibody molecule) can activate many different clones, whose surface receptors have different affinities for a given determinant. Antibodies include protein chains that can be divided into constant and variable parts. The antigen-binding portion of an antibody molecule is formed by two variable portions.

Affinity between antibody and antigen increases (affinity maturing) during the process of an immune response. It happens because the antigen "chooses" to reproduce those specific cells having genetic mutations that promote the establishment of bonds between antigen and antibody. It has been experimentally demonstrated that mutations occur in the genes that encode the variable parts of protein chains of the antibody. Moreover, the presence of a mechanism for increasing mutation frequency has been suggested. Thus, this schema of a lymphocyte learning to produce an antibody with high affinity for an antigen is a variant of Darwin's microcosm (see below). Cells of the immune "ecosystem" are subjected to variability and selection. The most highly adapted survive during selection. But "fitness" means bonding between antibody and antigen in each particular case.

Perhaps such learning by the immune system is the only experimentally demonstrated example of the generation of a new idea. This example is a special case of more complex learning processes capable of occurring in neural networks and is based on the concept that the intensity of a random search depend on the intensity of its initiating signal. In the immune system, mutations become more local and specific during affinity maturation, and should be considered to be a form of random search. The measure of discrepancy between the antibody receptor and the antigen is the initiating signal for this mutational random search. High antibody affinity for an antigen leads to the disappearance of this initiating signal. Obviously, it is possible to consider the above-described mechanism as one that limits the region of a random search in the state space of the immune system

It is currently accepted that immune memory is a result of differentiation of predecessor cells into memory cells, i.e., cells that do not produce antibodies themselves but can be easily transformed into effector cells during second contact with the same antigen. Memory cells are produced in large quantities. The increase in their number as a result of the division of precursor cells is referred to as "clone spreading." Memory cells circulate constantly between blood and lymph and can live for many months, and even years, without dividing. In contrast, the predecessor cells and the effector cells that mount an immune response die quite rapidly. There is evidence to suggest that this is only one of the many mechanisms providing immune memory. But before analyzing such mechanisms, let us consider idiotype-antiidiotype interactions.

Idiotypes are the antigen determinants of the antigen-binding portion of the antibody. Each antigen-binding region has a characteristic idiotype. Antibodies to idiotype regions- are called *antiidiotypic* antibodies. In turn, these antiidiotypic antibodies have their own idiotypes, to which antibodies could also be produced, and so forth. The main conclusion made on the basis of these facts is the potential existence of a complex interactive network of idiotype-antiidiotype types. Both T- and B-cell classes of lymphocytes participate in this interaction, since some idiotypes might be common to both of them.

Let us consider idiotype-antiidiotype interactions from the perspective of the OCS concept. It is possible to draw an analogy with the neural networks discussed in previous chapters.

First, an antiidiotypic antibody (or several antibodies) can prevent the binding of an antigen with its receptor. This mechanism may be considered as an analog of presynaptic inhibition, whose role has been analyzed in detail. Obviously, this may happen if such an antiidiotypic antibody does not exactly mimic an antigenic determinant, and does not activate a target cell (see the following analogy).

Second, the antiidiotypic antibody can mimic an antigenic determinant. In other words, it may function as an internal model of a particular antigen. In immunology, such antiidiotypic antibodies are called "internal antigen copies," and their appearance in the immune system has repeatedly been demonstrated. Such internal models and their antigens interact with a receptor on a parity basis.

Third, idiotype-antiidiotype interactions do not exhaust all the potential molecular mechanisms for interactions between immune cells. Different types of immune cells exchange information *via* numerous other signal molecules, but at present scientists know little about the details of these interactions.

Thus, there is a striking similarity between the immune and nervous system OCSs. The single difference is temporal, in that the immune OCS functions significantly more slowly. But just as in the nervous system, the immune system is capable of performing complex processes that result in optimal filtration. The following discourse stems from this similarity in the features of these two systems:

1. The immune system memorizes not only the antigen but also the whole antigen context, i.e., any correlations between different antigens. This endows it with the ability to respond at once with a set or pattern of antibodies corresponding to the antigens of the foreign agent, e.g., bacteria, during a relapsing infection. Obviously, this is one of the major advantages that the organism has during repeated contact with an infectious agent. For example, it is easy to imagine a situation whereby additional types of antigens appear, as the result of bacterial lysis, during the subsequent stages of the destruction of the infectious microorganism. If the organism is able to respond by utilizing an antibody pattern, it will be ready for the appearance of these "new" antigens in the future. It is also possible to suggest that not only is the antigen context memorized, but that the context of the concomitant presence of other signaling molecules is also memorized. Therefore, the latter context is also

capable of inducing an immune response, and a more rational and comprehensive explanation for neuroimmunomodulation phenomena, which are also being intensively investigated at present (see Spector 1987), might regard them as a result of such immune capabilities. For instance, an olfactory stimulus that was previously combined with the systemic presentation of a specific antigen becomes capable of evoking an immune response if the olfactory stimulus is introduced to the organism in the future.

2. The basis for immune memory includes not only memory cells, but also internal antigen models, as described above. The presence of the latter facilitates the optimization of the process of information storage. The establishment of a *functional coupling* between cells producing an internal antigen model (antiidiotype) and cells producing its antibody makes such a memory process very reliable with minimum material and energy expenditure. Such coupling is no different from the positive feedback phenomena between two cell types. The existence of "natural" antibodies, i.e., antibodies circulating normally in the blood, is inextricably associated with the functioning of such generator memory mechanisms. It is easy to imagine that the mechanism of control for the function of this ongoing memory task operate by means of constant suppression of antibody synthesis to maintain antibody production at a low level. The production of numerous populations of memory cells for each antigen is a wasteful process for the organism, and at the same time the production of too few of these cells is dangerous because their random death would deprive the organism of its cellular and humoral defenses. Precursor cells are also not a very reliable memory system for storing information. Any antigen "resembling" the internal antigen model of the precursor cell and having a weak but adequate affinity for its receptors (to induce an immune response and, consequently, mutations) can easily retune these cells.

Given the suggestion of the existence of the above-mentioned mechanism for memorizing a complex antigen context, it is possible to hypothesize another explanation for the cessation of humoral responses to an antigen. This process proceeds according to the scheme of control (1) by utilizing mismatches (with a desired result when the concentration of a particular antigen decreases to the point where there is no future binding between the antigen and B-cell receptors), and (2) by the creation of a model of the controlled object-antigen environment (necessary for the process of optimal filtration). Such a conclusion enables us to recognize some more analogies between the immune and nervous systems.

The existing theoretical concepts of the function of the immune system are analogous to the reflex concept in neurobiology. The latter, as was seen in previous chapters, is not complete enough to describe the function of the nervous system. The immune system, like the nervous system, is an OCS that includes an internal model of the controlled object. The model also probably includes any corresponding object dynamics (i.e., dynamics of the immune response to the intruding antigen). It is the latter feature—the capacity of the immune system to determine the dynamics of antigen context—that enables the immune OCS to determine a constant antigen composition, and thus underlies the concept of immune tolerance. Like the nervous system, the immune system halts its antigen production in situations for which it is incapable of decreasing an initiating signal.

The foregoing theoretical description helps us to better understand the problem of immunization. Effective immunization implies that a subject receives an appropriate context of antigens and other signal molecules; the more complete the context, the better. Only in this situation is the immune system of an individual capable of creating an appropriately effective internal dynamic model of the antigen context. The fact that a partial presentation of given antigen context for immunization (a dominant contemporary immunization strategy) is quite effective, in most cases, simply means that the immune system is constructed in a very clever manner. In fact, this means that the immune system is capable of solving complex puzzles. It is capable of mounting an appropriately specific response in the face of contextual poverty, in a new and more complex situation, after being trained to correctly respond to a much simpler stimulus. Obviously, a more complete and rich antigen context should be more effective in cases of immunodeficiency caused, for instance, by partial destruction of the system during the aging process or HIV infection. The immune system, like any other biological system, possesses holographic properties and has a decreased resolution power after partial destruction. Therefore, an appreciation of the immune system as a network system should be very fruitful for pursuing the development of effective vaccines in the future.

By tracing the analogies between the immune and nervous systems, it is possible to examine anew the problem of forming connections in the nervous system. The problem of the recognition of numerous molecular structures is effectively solved in the immune system. There are sufficient reasons to suggest that this invention of nature is also effectively used by the nervous system. It is enough to recollect the existence of mediators, more complex signal molecules (for instance, nerve growth factor), and their corresponding membrane receptors. If interactions of the ligand-receptor type underlie the basis for all complex connections in the nervous system, then the number of ligands and corresponding receptors must not be smaller than the types of connections between neurons. It is not possible to perform a detailed investigation of the whole variety of ligand-receptor interactions in the nervous system, but contemporary neurobiology is proceeding exactly in this way. It is expedient to investigate only the principles of this interaction, i.e., to carry out investigations of this problem from the perspective of the OCS concept.

It is obvious that the system controlling the structural framework of the nervous system must be an OCS subsystem and must interact with other brain OCS subsystems. Nerve growth factor is a typical example of an initiating signal for a structural neuronal automatism. A target cell produces nerve growth factor until contact between the growing cell (the cell that undergoes growth under the influence of the nerve growth factor) and the target cell is established (see, for example, Moffett et al 1996). Obviously, in the case of growth factors, we are dealing with a typical example of structural automatisms and structural optimization.

Finally, if ligand-receptor interactions are widely used in the nervous system, then the latter must be effectively isolated from the immune system. Otherwise, the immune system would work against the nervous system by preventing the appropriate establishment of accurate connections. As is well known, the blood-brain barrier plays this role, and its impairment results in serious dysfunction. Furthermore, such a bidirectional barrier is necessary for the protection of the immune system from various substances produced in the brain during intense learning processes. If the suggestion that the neuronal genome is involved in the process of active searching during learning is correct, then numerous molecules foreign to the immune system should appear in the brain during this process.

14.2.2 Intracellular Systems

The universal presence of protein molecules in the living organism, both for building different subsystems and for signal transmission between them, prompts one to consider describing the work of an intracellular OCS and its subsystems as automatisms of protein regulation. A large number of cell types are present in any highly-developed multicellular organism. Taking into account the fact that all cells have the same genotype, it is logical to consider all cell types as different states of one generalized cell-system corresponding to the genotype of a given organism. From this point of view, the process of cell differentiation in ontogenesis could be considered to be the control of cellular systems, which consists of writing in their memory the information about specific cell types.

Molecular biologists concluded long ago that a change in only one gene is not enough to cause a particular result. Changes in a system of genes interacting with one another are necessary. It is well known that interactions between genes are diverse and complex. That notwithstanding, the major features of their functional relationships are the same as the interactions between neurons: "excitation" and "inhibition."

Consequently, just as with neural networks, it is possible to describe functional networks of interacting genes. Ensembles of simultaneously active genes determine intracellular protein compositions and patterns of "control signals," in the form of the vast varieties of substances produced by the cell. Presently, the specific mechanisms and modulators involved in such interactions are not known in detail, but it is possible to suggest with high probability that protein molecules are the principal modulators and that mutual gene influence is performed by the reordering of protein synthesis. For instance, enzyme concentrations can play the same role in gene networks as synaptic weights play in neural networks. As a rule, enzymes are proteins. Consequently, it is easy to construct complex modulating schemes resembling presynaptic inhibition. The existence of feedback connections in gene networks and the creation of internal models of a controlled object are not excluded in this conceptual paradigm.

The cell as a whole should be considered as a complex hierarchical OCS having a corresponding hierarchy of states and of time scales of changes in its parameters. For all this, the upper hierarchical level is determined by a particular set of cell types in the organism and by a description of hierarchical states that contains not only a description of the states of genome activity, but also all sets of the subsystems for intracellular regulation, ferment content, receptor pattern, membrane channels, etc. It would be an unacceptable oversimplification to consider the cell as a system possessing only a set of static states and concomitantly void of its own dynamics. Such dynamics are well delineated in most cases. It is thus possible to identify a model of controlled object dynamics in intracellular OCSs. Examples of such models are described below. As stated earlier, the cellular OCS is hierarchical, and automatisms of different levels within it should be identifiable. Let us recollect the role of the enzyme in converting one substance to another. Most frequently, the product of the reaction can influence the activity of the enzyme by changing its conformation and inactivating it. Thus, a typical system of control consistent with the mismatch concept can occur in this case. Internal models appear in systems controlling automatisms at higher levels as well, when the necessity for predicting object behavior arises. Let us consider two cellular automatisms: "export production" of proteins and the pacemaker activity of a neuron.

Recently, experimental data have appeared that point to the important role of two membrane proteins in the cellular regulation of "export production" of proteins (Kulberg et al 1987): (1) receptors recognizing the production of a given protein, and (2) antireceptors imitating the structure of that particular protein. The classical schema of feedback control for this process takes into account only the inhibition of synthesis that takes place during the binding of membrane receptors with protein molecules. It is obvious that antireceptors may compete with protein molecules on a parity basis and the analogous description presented in the previous section regarding the role of antiidiotypes in immune response regulation can be applied to cell function in protein synthesis. In other words, the production of antireceptors testifies to the presence of feedback through an internal model in the cellular system controlling the external concentrations of proteins synthesized by the cell. This internal model therefore describes dynamic changes in external protein concentration.

One possible mechanism of chemically induced pacemaker activity has been described in detail for spinal neurons in the lamprey (Grillner et al. 1986). Ion channels modulated by NMDA receptors are responsible for the depolarizing phase. The distinctive characteristics of these channels are that (1) their conductivity depends on the extracellular concentration of magnesium ions, and (2) between -60 and -40 mV there is a negative slope on the current-voltage curve of neuron response following NMDA application. Therefore, in response to the application of a constant mediator concentration, a wave of depolarization appears. During this wave, calcium enters the cell and activates calcium-dependent potassium channels. The latter phenomenon leads to cell repolarization.

In connection with the previous paragraph, let us recollect the intracellular system of calcium regulation. Calcium is released from its intracellular stores under the influence of different factors, for instance, inositol phosphate or the entrance into the cell of extracellular calcium (Berridge and Irvine 1989). Different types of intracellular calcium oscillations have been described. It is significant that some oscillations disappear just after the removal of extracellular calcium, while others are preserved for a long time. It is possible to suggest that the intracellular system of calcium regulation models (at least partly) the behavior of the neuronal membrane and hence of the neuronal environment, and that the intracellular system of calcium regulation predicts the occurrence of calcium entry into the cell. Extracellular calcium and calcium liberated from intracellular stores equally influence intracellular systems: both calcium types effectively exert an influence on calcium-dependent potassium channels. Therefore, oscillations in intracellular calcium concentrations should be considered as a reflection of internal model activity. The presence of such an internal model provides the system with the property of a highly reliable pacemaker (for instance, low dependence on the concentration of external calcium, which undergoes seasonal changes in some animal species).

Therefore, in each case in which we have identified in a molecular network a molecule that can be considered to be an internal copy of some signal molecule, or a molecule that can bind with corresponding receptors or enzymes, it is possible to posit the existence of an internal molecular representation of the corresponding molecular object and to discuss its dynamics.

In contemporary molecular and cell biology, the majority of research still proceeds according to a simple schema of stimulus-reaction reminiscent of the stage of reflex theory in neurobiology. Neurobiology, however, is about to leave this "reflex" stage and move toward a more profound understanding of brain phenomena based on the concepts described in previous chapters. It would be very beneficial for molecular biology and other biological sciences to make some predictions about their future development based on the experiences accumulated in neurobiology

14.3 Evolution and Learning Processes

Let us remember the mechanism of evolutionary processes proposed by Charles Darwin, which any student who has studied biology in the twentieth century has learned. It is based on variability and natural selection. If a "nonoptimal" individual (for a given condition) appears as a result of mutations, it is highly probable that this individual will soon die. It follows from the previous chapters that the evolutionary process has to be considered as a learning process, and that Darwin's schema is the most simple and primitive of the possible mechanisms of learning. It must have played a significant role only during the early stages of evolution. To demonstrate this, we must return to the mechanism of random search learning (Chapter 6).

This type of learning embodies the idea of learning without a teacher and needs a gradient of an initiating signal. At first glance, Darwin's schema possesses these features. However, it has already been revealed that when the number of changeable parameters in a functional system is large, this learning mechanism becomes very slow and ineffective if *any* parameter can be changed during a random search. There should therefore be mechanisms for accelerating evolutionary processes. Discussions of possibilities for the acceleration of evolution have a long-standing history, because it is difficult to believe that the most complex of organisms existing at the present time developed according to such an ineffective learning scheme. It is intuitively clear that evolutionary mechanisms should be improved and become more complex with the development of organisms with progressively greater complexity. But the nature of these mechanisms in higher animals is practically unknown.

Recent investigations in the field of mutagenesis have revealed a very interesting peculiarity: the intensity of mutations in separate genome loci may increase by several orders (Alberts et al 1994) under the influence of different factors (including biochemical). We already saw how this mechanism is used in the immune system (see Section 14.2.1). Localized mutagenesis means that a random search is limited to a very small subspace of the state space of the system. Obviously, it is natural to propose that a localized random search can work in evolution, because it accelerates this type of learning significantly. This also means that corresponding automatisms have to appear in evolution. In this case, the state space of the system has to be understood as being a state space of an organism, and a spatially limited search is performed in the area of space that describes a specific feature, or features. Such a suggestion means that each type of initiating signal received by an organism has to influence a specific region of the genome, and the corresponding molecular mechanisms for precisely and accurately orchestrating such an influence must exist. Otherwise, the evolutionary process would be excessively prolonged, or even impossible in cases of complex life forms. A significant evolutionary acceleration could take place by moving against the gradient of a localized mutation intensity, and hence against the intensity of its corresponding initiating signals, i.e., moving towards adaptation.

Let us consider several possibilities:

1. Evolution of ecological associations or, in particular, species. This process was described above and is performed by using localized mutations. The intensity of the corresponding initiating signals decreases in individuals mutated in the direction of adaptation to new conditions and, correspondingly, the intensity of mutation decreases so that their genome achieves a new state as though it were "frozen." The exchange of genetic material within species may promote the spreading of new adaptive mechanisms. Obviously, this effect is increased by different factors associated with a dependence of individual behavior on the signal intensity that is initiating mutations.

2. The development and "perfection" of a separate individual. If the individual organism is viewed as an association of cells, then the above-described schema may be applied to an individual organism. The evolutionary process then becomes directed toward the organization of genetic information exchange between cells. Whether or not such a process exists is presently unknown. It is not difficult, however, to imagine that such processes might be indirect and mediated by different signal molecules. Such a process, to a certain extent, may be conceptualized as a "descent of an automatism." It is simpler to duplicate any advantageous solution obtained earlier in the course of evolution than to search for new solutions in each cell independently.

3. Evolution of the neuronal structures of an individual's brain. If a powerful immune system is used in the organism for the recognition and destruction of mutant cells, then the "laws" of the immune system will be of lesser value in the brain, since the immune system is separated from the brain by the blood-brain barrier. In this regard, it is possible to significantly expand the set of degrees of freedom and provide greater variability for neurons and glial cells. Consequently, the control of mutagenesis in brain cells can play a significant role in learning processes.

It follows from the above discussion that the use of intensity to control random changes in a given control system is the most interesting and important biological mechanism on which learning processes and evolution are built.

That is why the investigation of such mechanisms is most important and potentially fruitful for understanding how any biological system learns, for the control of learning, and also for the creation of artificial learning OCSs.

An understanding of the mechanisms of random changes permits one to make several important conclusions. First, a global minimum on an error surface (the surface of an initiating signal) cannot be reached instantaneously. Therefore, some intermediate stages (transitional stages) in the evolution of a given species should exist. But any intermediate decisions must permit an animal to survive. Second, a global minimum state is not always attained. However, in the case of older species that match their ecological niches well, the majority of their evolutionary decisions are optimized in such a way that they achieve their global minima on the error surface.

One more aspect of evolutionary processes deserves attention. It is the creation of an internal model of the controlled object. This necessity is a logical extension of the features of optimality present in any OCS. We can imagine this process in the following way.

For a control system not receiving afferent signals—neither initiating nor informational—the controlled object exerts only physical influences on the system. Any interaction with the object is described solely by physical laws; it is nonbiological in nature.

An OCS receiving initiating signals (which must be minimized) represents an evolutionary expansion based upon three emerging features. Let us consider the example of the movement of an evolutionarily primitive organism away from the source of a sound that serves as an initiating signal, by means of a primitive effector organ such as a tail. The afferentation from the effector organ could be of three types:

1. An *absence* of any afferentation. In this condition, a model of effector organ movement must be present in the OCS to at least provide cyclic movement control. In such a case, the presence of the model may be considered to be a necessity for the program control to be of sufficient complexity. Moreover, the development of such a model can only take place according to the principle of "learning without a teacher." For this reason, it is possible to create only primitive models acceptable for primitive effectors such as cilia, flagella, etc., under such circumstances.
2. *Primitive afferentation*, terminal sensors, and pain receptors. In this evolutionary scenario, the richness of afferent information is not sufficient for obtaining reliable information about the current state of the controlled object. Signals received by an OCS from its effector organ have an initiating character, informing the OCS of any potentially aversive state in which a given organ as a controlled object might find itself.
3. *Developed afferentation*, including initiating signals as well as sensory signals, the context of which is informational. It should be noted that

signals containing an informational context are a new kind of mismatch signal, which appears as a result of a significant increase in the resolving power of the system producing these signals.

The visual organ is a typical example of the third type. In the presence of a small number of photoreceptors, the reliability of transmitted information is low, and therefore this organ can only distinguish between light and dark. A signal about transition from light to darkness (or vice versa) may, for instance, serve as a signal of danger and initiate escape reactions. Qualitatively different situations appear in the presence of a more extensively developed visual organ (eye) in higher invertebrates and in vertebrates, providing more detailed information about the smallest of changes occurring in the environment. In the case of acquired afferentation, it is simpler to create an internal model, because the controlled object itself plays the role of a "teacher" while the model only has to mimic the information coming from the object.

The mechanism of the gradual transformational exchange of initiating inputs for informational inputs is very interesting. The necessity of such a transition is quite clear. It follows from the need to minimize the sum of all initiating signals. At the same time, this transition heralds a change from simple feedback control to program control. The correctness of this conclusion—not obvious at first sight—is revealed in the following example.

Let us consider the situation in which an ambulating animal's limb encounters an obstacle in its trajectory. As a result of missing an anticipated step, the corresponding initiating signal induces a correction of movement. This is a typical example of the working of negative feedback. However, the alternative situation is possible—the animal may see the obstacle in time and try to avoid it. In such a case, movement control becomes the programmed type, and, moreover, contextual information of a visual nature is used as an input. It is clear that information processing is more complex in the second case. This is the price for the system optimality feature. It is possible to cite numerous examples demonstrating that the development of afferentation takes place in such a way. For instance, in phylogenesis and ontogenesis, the mossy fiber system in the cerebellum appears and gradually becomes more complex after the development of the climbing fiber system.

Given this general tendency for developing afferentation and an internal model of the controlled object, it is possible to draw conclusions regarding the functional complexity of the brain and other complex non-neuronal OCSs. In lower animals, behavior primarily has an instinctive and reflexive character. Learning occurs by the establishment of conditioned reflexes—the creation of models of initiating signals. In higher animals, the gradual formation of the cortical model of the environment takes place, i.e., the cortical model of a more global informational context. The most interesting future prospect is to understand the new neurobiological mechanisms for the intensification of learning processes in the human cortex that were created by nature.

14.4 Self-Applicability of the Theory and Its Application to Other Sciences

The problem of constructing a theory of brain function is quite specific. The brain is the principal instrument for the development of any science, including neurobiology itself. Therefore, it is necessary to demonstrate the applicability of the brain theory to any science, including the theory itself. This conclusion was the reason that serious doubts were cast upon even the faintest possibility of fully comprehending the brain, during the period of powerful developments in the field of artificial intelligence in the 1960s and '70s. The potential for completely understanding the brain was compared with the possibility of lifting oneself by pulling up one's hair.

Such doubts were not completely unfounded, because it was known in mathematics at that time that self-applicability frequently led to the appearance of contradictions and paradoxes. However, alongside this were several known cases in which these paradoxes were readily resolved. The problem of constructing a self-reproducing automaton (solved successfully by von Neumann in 1951) provides a good example of such a solution. The reproduction problem was for a long time one of the cardinal challenges in biology. It turned out that the solution of the paradox of an infinite sequence of germs enclosed within each other is the replacement of "reduced copies" by the code for the organism (automaton, program text). Analogously, it is unnecessary to have "small copies" of neurobiology, physics, chemistry, mathematics, and other sciences in brain theory. It is only necessary to find an effective way to encode them.

What is common in all sciences? They are all models of different environmental aspects that are important for man's life. The term "model," used in the most general sense, is understood as a system of knowledge about the corresponding object. This knowledge is divided into two major functional subsystems:

1. "Passive" knowledge, i.e., knowledge about how an object behaves in response to a given influence upon it
2. "Active" knowledge, i.e., knowledge about how we should influence the object in order for it to behave as we desire

Each of us knows that the task of mastering "active" knowledge is more difficult than the task of mastering "passive" knowledge. Illustrative examples of how the brain performs these tasks were considered in previous chapters.

Let us try to trace the development and formalization of ideas regarding the nature of the environment that have enabled scientists to describe brain function, i.e., the system in which these ideas appeared. Such an analysis is useful because brain research and any sufficiently formalized science (for instance, mathematics) are systems of knowledge presentation that reflect the environment. That is why the principles of their structural organization must be similar to a certain degree.

Different fields of mathematics were developed as a result of distinguishing some particular aspect of the environment that could be described quantitatively. It is possible to distinguish three major aspects of this sort:

1. The system of cause-effect connections (mathematical logic)
2. Object reiteration, when abstracting from those descriptive aspects in which there is a difference (quantitative characteristics expressed by numbers)
3. Spatial object form, a spatial analog of cause-effect connections (geometry, more exactly, topology)

Mathematical logic might be perceived as the first real model of thinking. At the same time, as a branch of science, it primarily reflects a cause-effect organization of the environment. It is interesting to note that the major rule of logical conclusion—*modus ponens*—could be conceptualized as the simplest function performed by the neuron: if there is a specific excitatory input signal, then an output signal will appear.

The axiomatic principle of construction of formal theories was a generalization and development of the theory of logical conclusion, as if a superstructure above mathematics (metamathematics) appeared that defines theories noncontradictory in nature. As a result, the thoroughness of different axiomatic theories was investigated. Further expansion in the field of mathematics, or fundamental concepts and mechanisms, underlay the basis of thinking processes, and led to the construction of a theory of formal grammar, languages, algorithms, and recursive functions. In general, an algorithm is understood as a description according to which a man or a machine performing a finite number of simple operations during a finite time can obtain a necessary result. The agreement of this approach with the theory of numbers (a number is used as a universal code) led to the appearance of the idea of computability and to the theory of computing machines beginning with the advent of Turing's machine.

The second most important mathematical branch originates from formalized concepts of the quantitative characteristics of the environment. The concept of a number, in turn, is based on the identity relationship between environmental objects. This identity is only obtained within a specific subset of object characteristics. At least one of these characteristics must be variable. Such a characteristic is usually the spatial position of the object. The concepts of number, arithmetical operation, and sets evolved in this way.

The third branch of mathematics—geometry—appeared as a result of a formalization of ideas regarding shapes of environmental objects, which one could interpret as a spatial analog of simple cause-effect relationships. (Topology is meant to be an abstraction from symbolic numeric characteristics.)

Subsequent mathematical developments are metaphorically comparable with a tree whose roots consist of the three previously enumerated origins. For instance, systems of differential and integral equations could be viewed as a special form of grammar in which logical and numerical origins are intermingled. It is interesting that geometrical concepts were also successfully utilized in the investigation of the behavior of such systems.

The three principal concepts of mathematical systems development described above are used in the current description of brain organization: *geometrical*—the structure of connections between neurons, membrane forms, and synaptic contacts; *quantitative*—the frequency and amplitude of electric signals, synaptic weights, electric membrane parameters, etc.; and *logical*—the functions performed by a specific neuronal system.

The problem of correspondence between structure and function appeared at the very beginning of the development of the brain sciences. Most of the ideas concerning the structural and functional organization of the brain and the construction of cybernetic apparatuses were proposed within the limits of this problem.

The current stage of scientific development is characterized by a narrow specialization of research. The profound interconnections among the sciences—not only among the branches of one science, but also among different sciences—become clear. The aforementioned specificity of the brain sciences—which may in fact be formally designated a "theory of neuronal self-learning systems of optimal control"—is the basis of hope for integrating the different sciences within this discipline. If one is to pursue this conceptual course, it becomes necessary to return to the idea of generalizing the theoretical concepts developed in this book to systems that have no neuronal structure, but are nevertheless constructed according to the principles of functional organization.

Some examples of the application of conceptual generalization to biological non-neuronal systems were discussed in the beginning of this chapter. Furthermore, there is the very attractive possibility of revealing the potential applicability of the conceptual theory outlined in this book to an ecological system, or even to human society, which undoubtedly is constructed according to the principles of functional organization. The theory also provides a means by which an artificial system having the functional abilities of the human brain might be constructed.

The social milieu that is human society presumably functions, on a conceptual basis, as a network computer consisting of a hierarchy in which each level creates a model of the behavior of its controlled objects, etc. In a society, for instance, it is possible to find numerous types of initiating and informational signals. Money is a typical example of a powerful initiating signal that each individual or segment of a society, such as business, is trying to maximize (to minimize a shortage of money) by executing learned automatisms. As was observed in neural hierarchical systems, an effective description of environmental cause-effect relationships requires a perpetual ascension, within the system, of its parameters' degree of abstraction. This phenomenon was probably the basis for the creation of language itself. Language became a universal tool for creating new abstractions because, after the creation of language, new abstractions could be constructed by implementing new symbolic concepts or *words*. From this point onward, the power of an intelligent biological network computer in the form of an individual or a society began to rely heavily upon external memory mechanisms, because its own capabilities for accumulating knowledge were limited by the life span of man. External memory was initially confined to the use of storytelling in the social group to which an individual belonged. Later, the social group acquired an unlimited ability to store information in a collective external memory with the creation of written language. From this point forward, the evolutionary process in human societies depended on the effectiveness of access to information contained within this collective memory.

Within the framework of these ideas, it is possible to consider any science as a specific form of social automatism. Like any automatism, it is a learning system and creates a model of the controlled object. For instance, any new scientific method can be considered as an analog of a new effector. Its further elaboration and mastering by others is a creation of a corresponding control system that possesses a corresponding model. Further optimization of this control system leads to an understanding of the limitations of the method, partial loss of interest in it, outflow of scientists from the field, etc.

It is not the goal of this book to present a deep and detailed discussion of the organization and control of science. But some intriguing ideas relating to the organization of science—for instance, neuroscience—deserve attention and immediately come to mind if science is considered as an automatism.

Like any other automatism, a science is controlled by its corresponding initiating signals and uses any available informational context to minimize these initiating signals. The question arises: Has society created appropriate initiating signals for scientists, or are these signals inadequate criteria for minimization, and have they led to the creation of a parasitic scientific automatism because the system responds to improper initiating signals? If the latter is the case, then the control of science is far from being optimal. Let us attempt to identify the fundamental initiating signal for a scientist. I think that the majority of scientists would agree that it is curiosity. This is the most powerful driving force for any real science. Curiosity has strong evolutionary roots: it is well known in behavioral sciences that there are individuals even among highly organized mammals whose strong curiosity can easily lead them into dangerous situations. Such individuals can be viewed as "explorers." We will not consider other initiating signals, for instance ambition, since they are usually realized through a different system of rewards. Other signals are most often secondary and, as a rule, are not present at the initiation of a research program.

Therefore, a control task for a society consists of creating a mechanism—an automatism—that financially supports curiosity. This would be an optimal solution from a theoretical perspective. The current grant system—one of the most efficient systems that was ever created to control science—was designed to function in this manner. However, the grant system quite often creates pure financial initiating signals for scientists. Curiosity has moved to second place. A strategy of doing what we *can* instead of doing what we *should*, what we are interested in, and what is best for understanding the phenomena of nature has become quite a widespread situation in contemporary science. Obviously, the substitution of money for curiosity, as an initiating signal, has not led to the creation of optimal control. Instead, it has created a set of new and unnatural initiating signals for science—such as the number of publications as a reflection of the quality of a research program, the number of years spent conducting research in a specific scientific field (called "experience"), even without advancing the science itself, etc. These newly "invented" initiating signals have begun to play a conservative role in preserving the existing model of science, instead of supporting modification of an automatism—a situation that merely serves to impede any meaningful scientific progress. It is hard to imagine, for example, Albert Einstein writing scientific articles in order to have a substantial enough number of them for promotion and tenure.

14.5 Future of Neurobiology for Physicists

Given the concepts developed in this book, it is possible to outline the future of neurobiology and the role of physics and mathematics in it. It is clear that the neurobiology of the future will become a highly technical discipline and draw heavily from the wealth of both theoretical and applied physics and mathematics. The calculus of variations, the theory of differential equations, functional analysis, and the theory of algorithms will routinely be utilized in neurobiology. Physicists know extremely well how to use differential equations to describe the behavior of a physical system, and how to solve variational problems of differing complexity. The brain is a physical system and, as we have seen, the behavior of its subsystems strongly depends on effective minimization of its corresponding initiating signals, i.e., its corresponding criteria. Any physicist will find this fact fascinating; the principle of least action is one of the central principles of modern physics.

It is difficult at present to predict which additional concepts from modern theoretical physics or mathematics will be successfully applied to and incorporated into the neurobiology of the future. But one fact is obvious. New theoretical abstractions will be necessary, and physicists and mathematicians are quite accomplished at creating formalized theories. Unless these new theoretical abstractions are forthcoming, it will not be possible to adequately address the following very important theoretical issue: how detailed can our model of the brain be? From a theoretical perspective, the process of brain cognition resembles a situation in which one learning system (man) is trying to create a model of another learning system. Physicists know that there are objects whose states cannot be determined completely (for instance, Heisenberg's uncertainty principle). It is highly probable that the brain possesses the same features, and that its states also cannot be completely determined or predicted. The presence of numerous hidden parameters that an observer (a scientist) cannot measure may be one of the reasons for this. We also do not know yet which theoretical concepts possess the necessary functional completeness for describing the behavior of the brain. For instance, does the theory of algorithms possess this power? Is it possible to completely describe the behavior of the brain with an algorithm, or does some other concept in addition to algorithms have to be used for this description? Therefore, an expansion of the use of physics and mathematics to describe biological neural network computers (any biological network computer) will ultimately result in the creation of a more effective and more functionally complete language for constructing a model of the nervous system. The formulation of theoretical problems of this type is not commonplace in neurobiology, but their usefulness for neurobiology has already been demonstrated. For instance, the network computational principle came from technical fields (see Kolmogorov's theorem and backpropagation neural networks).

Will all this mean that it will become easier to conduct theoretical and experimental research in the neurosciences? Definitely not, because a great deal of new knowledge has to be acquired by a scientist in order to conduct successful research. But eventually research can become much more goal-oriented, such that the component of random searching—an inherent feature of any evolving research program or *any* cognition—will be significantly reduced and limited to the confines of a more scientifically productive portion of a given state space.

Let us go back to the network computational principle. It is clear that accentuating the computational interpretation of the function of biological neural networks has several profound consequences for the neurosciences. One of them was already mentioned above. It is the possibility of applying the power of physics, mathematics, and control theory to the analysis of biological neural network computers. Mathematics has already proven itself to be powerful enough to describe different controlling systems, including artificial neural networks.

However, in the contemporary disciplines of artificial neural networks neurocomputing, and computational neuroscience, only the simplest of networks, which usually incorporate only symmetric connections, have been developed for an analysis of the nervous system. Much more work must be done to develop more powerful methods that utilize mathematical descriptions of more complex neural networks. Modern analytical mathematical descriptions are effective only in cases of simple neural networks. Concerning computer simulations, the power of solving a set of numeric equations that describe the function of a complex network has quite often been applied to the well-known *toy* problem (see below). Frequent use of such methods for solving the toy problem could potentially discredit this computational approach.

However, at this stage in the development of contemporary neuroscience, the *toy* problem can be justified. Scientists must continue to invest a great deal of energy to develop even the simplest of network simulation programs. This is why scientists often make countless modifications that usually do not result in any qualitatively new functional results when a network simulation program is completed and the first simulation results are obtained (This is exactly what the toy problem is). Moreover, in some cases the thinking of investigators is also quite simplistic: (1) To create a network with an architecture similar to an actual biological neural network or, more precisely, similar to what we know presently about biological neural networks, and (2) to include in the network as many neurons as possible with the hope or the assumption that doing this will somehow bring about a new quality to the research. However, no one has succeeded in obtaining any such new qualities by using an approach based on these assumptions. This situation reminds one of the current state of the experimental discipline of systems in neurobiology. For some unknown reason, it is also very popular to perform simultaneous multiple-unit recording from unidentified neurons in an awake animal performing specific behavioral tasks. Clearly, it is necessary to consider the toy problem as a natural inherent feature of the cognition process. As mentioned, the latter includes random searching as a part of its learning strategy. Any discipline of science must eventually face this dilemma because of certain limitations it reaches at each discrete stage of its development. Interestingly, one can conceptualize these limitations from the perspective of a research program conducted by any individual scientist. At each stage of development, a scientist first models his own abilities to describe the object being investigated in his own research program, i.e., his own mental abilities.

As one can surmise from this book, a more fruitful and effective experimental and theoretical approach might be to start from the groundwork of conceptual and functional modeling, i.e., utilize the approach of analyzing which architectural and/or functional modules, containing which computational abilities, have to be included in a particular controlling system in order for it to perform a specific function for an individual organism. The creation of a conceptual description of a particular neural network to be investigated might be considered as the initial stage of functional modeling, when its major minimization criteria are determined. The finding of these criteria can be one of the most difficult aspects of this approach. The explanation for this approach is clear. It arises from the fact that the same *type* of computation can be performed by networks having different architectures and elemental properties. If the computational abilities of each individual functional OCS are determined during functional modeling, then it becomes possible to show what types of networks consisting of unique and/or specific functional components are capable of performing any necessary computations. In this case, we can think of it in the following way: Why bother making computer simulations that demonstrate that a given network can compute a specific function, if it is already known that networks composed of an architecture identical to the network being simulated include this same function in their class of computable functions?

Functional modeling can help us avoid making costly and time-consuming errors. For example, if phenomenological modeling (modeling that mimics the behavior of the system) does not account for some unknown parameter(s) of the system being investigated, then it is most probable that a wrong conclusion about the computational abilities of the modeled network will eventually be made. For example, simulations developed for explaining locomotion that lack complete knowledge of the parameters necessary for such movements will result in a situation analogous to the case of a toy dog that mimics the walking movements of a real dog but does not meaningfully elucidate the principles of neural control of this motion.

The complexity of the computational problem that exists for the functional approach to the nervous system is the primary reason that the present book does not analyze the numerous results of computer network simulations. In many cases, the simulated networks currently available are very simple, and it is obvious that the function of their biological prototype cannot be reduced to such simple simulations. However, several unique and surprising features of simulated networks have been demonstrated, and some of these were mentioned in the previous sections. In general, computer network simulations may for quite some time be the only way to demonstrate the computational abilities of a specific material network.

With the functional approach, the complexity of computational problems can lead to a situation in which it is unclear whether it is necessary to obtain a time-consuming mathematical solution, or whether some other analytical approaches to the mathematically formulated problem might be more appropriate. It is well known in physics that, most often, a mathematical formulation of a physical problem is the most important stage of a research program, and that it is usually possible to determine a surprising number of properties of a system almost immediately after such a formulation is made. Obviously, a functional description can be much more compact than any other description of the object under investigation. For instance, a description of any computer program at a functional level (a description of what each block of the program has to compute) is much less voluminous than the actual listing of commands for the program itself. From this point of view, it also becomes clear that if all the synaptic connections in the brain were known, such information would most probably be useless because of its unbelievably huge volume.

Obviously, the concept of computation also changes our understanding of the relationships between anatomical structure and neurological function. The major problem in the relationship between these two subjects can only be solved by performing an analysis of a network's computational abilities, thus revealig the connections among function, computation, and structure. In this field, almost all of the problems and questions to be addressed by researchers should be reformulated. For instance, a neuron can be viewed as a universal computing device (a universal processor), in which a relatively simple initiating signal leads to a realization of a mathematical function. In this way, it is also possible to understand the developmental mechanisms of a network. Clearly, any network has to be genetically predetermined to compute specific classes of functions. Sources of initiating and informational signals (detectors) for each level also have to appear during specific stages of ontogenesis. Otherwise, the system will not be capable of learning, will not create a necessary model of controlled object behavior, and will not have the ability to perform the proper control tasks. If there were no initial structural approximations of the network system by way of genetic predetermination of the networks used to compute specific classes of functions, the process of learning would be much longer, and a scenario in which learning does not result in the creation of any essential automatisms in a lifetime would be normal. The existence of complex biological systems would therefore be impossible because this would be analogous to finding a new solution without any previous knowledge of how new solutions have been attained in the past, i.e., without a knowledge of how it was done by nature in evolution. One of the best-known examples of such an initial structural approximation, of mature structural optimization, in a network system, is the cerebellum. Hard-wiring of its network occurs after birth when lower automatisms have already formed. In the initial stages of cerebellar wiring, several climbing fibers make connections with a single Purkinje cell. Later, only one climbing fiber input survives. It is likely that other climbing fibers, whose signals could not be minimized, were rejected. It is well known from control theory that an optimization of parameters is a much less complex computational problem than structural optimization, and it usually takes much less time.

It is also clear that some cherished biological notions have to be completely abandoned—for example, connectionism and the single-cell hypothesis. The first notion places greater importance on a network's parameters, while the second states that the neuron is the most important component in network functioning. It is clear from all the above that these points of view are complementary: the more powerful the network's elements (neurons), the simpler its schematic manifestation can be.

The relationships between the brain and the mind, self-awareness, and other psychological constructs can also be characterized as structure-function relationships. It is possible to understand this relationship only by revealing how a highly abstracted model of an individual and the environment is created in the brain and how the brain uses this model to perform computations.

In general, the above discourse will ultimately result in the neurosciences becoming more oriented toward revealing principles of function, and the pointless collection of data will stop. Simply accumulating information cannot help us to understand any new principles, because there are countless numbers of neuronal geometries, nervous systems, types of ion channels, and other molecular mechanisms.

14.6 Artificial Intelligence and Future Neurocomputers

In order to understand what neurocomputing will need in the future, let us briefly summarize the principles of construction of the hierarchical biological neural network computers described in previous chapters. All hierarchical levels are constructed according to the same functional schema. Each controlling system is a learning system and has the same functional components (a substrate containing a control law and a substrate that models the behavior of the object) and the same types of afferent inputs (initiating, mismatch, and information). The network hierarchy is created in such a way that a lower control system becomes the controlled object for a higher, more sophisticated control system. Naturally, the hierarchy leads to a generalization of its encoded parameters and, consequently, the working space for each consecutive hierarchical level becomes more and more abstract. This is why each hierarchical level utilizes specific detectors that match its control needs, in order to obtain any necessary information pertinent to the current state of the object in its surrounding environment. Any movement toward higher hierarchical levels is also accompanied by an increase in the duration of ascending initiating signals. This is why the higher the control level, the longer the stored history accessible in the form of memory. The ability to store a long history was a precondition for the development of more and more complex and rapid forms of learning, such as operant learning; it also made it possible for the system to perform complex multistep computational algorithms that require the ability to memorize the results of any intermediate computations. At its highest levels, the capacity of the brain to utilize fast forms of learning based on complex molecular, cellular, and network mechanisms made possible the creation of detailed cause-and-effect models of the surrounding environment.

The power of such a computer depends on the computational abilities of each of its levels, its speed, its ability to learn, and its memory capacity, because each level has to store its own control law and a model of the behavior of its controlled object. The more complex the controlled object, the more sophisticated the network has to be. The process of learning in such a system will eventually involve all of its hierarchical levels. Learning first occurs at higher levels, and then the process leading to the descent of an automatism is activated as the lower levels within the hierarchy learn to remove any arriving initiating signals.

Obviously, if we could create such artificial networks, they would be neurocomputers utilizing the principles of construction of biological neural networks. From the above discourse, we can now attempt to understand what is missing in modern neurocomputing.

Modern neurocomputing only utilizes relatively simple neural networks for which mathematical analyses are possible. These simple networks have, as a rule, symmetrical connections, and a great deal of attention is focused on the development of effective learning algorithms because learning usually starts from scratch. This approach is adequate for simple computational problems. However, in cases of complex computational problems, for which algorithms of learning, if any, are not known, it would be most beneficial if an initial structural approximation of the system in question were available for subsequent learning sessions. Such "genetic" approximations can significantly accelerate the process of learning because the network is designed to compute a specific class of functions.

Hierarchical networks in the biological sense are not used in neurocomputing. It is clear from previous chapters that one should not expect great high computational powers from nonhierarchical neural networks. Complex computational problems arise when the controlled object is hierarchical and has a low level of observability and controllability. This is the major reason why hierarchical biological control systems possess a model of object behavior. Therefore, future artificial neural networks have to become hierarchical, and each hierarchical level has to have an internal model of its corresponding controlled object.

The technological problems that have to be overcome in order to create artificial neurocomputers are enormous. First, it is necessary to develop a new set of elements—*microchips*—that possess the principal features of biological elements—*neurons*. This does not mean that such elements have to completely mimic real neurons. These new devices do not necessarily have to be purely electronic devices either. They can be of a hybrid type that incorporates optical or chemical processes. More attention must be given to the learning principles used by such elements, i.e., to developing new technologies that can facilitate the implementation of different principles for creating long-lasting modifications to an element's parameters. Second, the optimal architecture of such computers can be elaborated only by performing a functional analysis of the controlled system (controlled process) and by finding a good initial structural approximation of the system being investigated. Identifying the initial structural approximation of such systems can be likened to the process of biological evolution. Clearly, if one attempts to invent rules such as those occurring in nature, it may take a remarkably long time to do so. The easiest way to acquire the knowledge necessary to devise an initial approximation of a network system is to discover how it was done by nature. Through evolution, nature has produced these kinds of adaptive results in its laboratory, having had virtually unlimited resources and time.

This new technology, although likely to be very expensive and time-consuming in the beginning, will without any doubt be most beneficial for mankind. Artificial network computers utilizing biological principles will be capable of predicting and controlling objects that have low levels of observability and controllability, such as social groups, the economy, the stock market, etc. There is no doubt that this will be the future of the computer sciences, because the problems of real life are much more complex than the problems that the contemporary generation of computers is capable of solving.

REFERENCES

Adams RD, Victor M. 1993. Principles of neurology. New York: McGraw-Hill Inc.

Alberts B, Bray D, Lewis J, Raff M, Roberts K, Watson JD. 1994. Molecular biology of the cell. 5 vol. New York: Garland Publishing Inc.

Alexander GE, Crutcher MD. 1990. Functional architecture of basal ganglia circuits: neural substrates of parallel processing. TINS 13:266–271.

Alexander GE, DeLong MR, Strick PL. 1986. Parallel organization of functionally segregated circuits linking basal ganglia and cortex. Ann Rev Neurosci 9:357–381.

Allen GI, Tsukahara N. 1974. Cerebro-cerebellar communication system. Physiol Rev 54:957–1006.

al-Tajir G, Starr MS. 1991. Anticonvulsant effect of striatal dopamine D2 receptor stimulation: dependence on cortical circuits? Neurosci 43:51–57.

Anden NE, Jukes MGM, Lundberg A, Viklicky L. 1964. A new spinal flexor reflex. Nature 202:1344–1345.

Andersson G, Armstrong DM. 1987. Complex spikes in Purkinje cells in the lateral vermis (b zone) of the cat cerebellum during locomotion. J Physiol 385:107–134.

Andy OJ, Jurko MF, Sias RF Jr. 1963. Subthalamotomy in the treatment of Parkinsonian tremor. J Neurosurg 20:860–870.

Anokhin PK. 1974. Biology and neurophysiology of the conditioned reflex and its role in adaptive behavior. New York: Pergamon Press.

Aoki N, Mori S. 1981. Locomotion elicited by pinna stimulation in the acute precollicular-postmammilary decerebrate cat. Brain Res 214:424–428.

Arbib MA. 1987. Brains, machines, and mathematics. New York: Springer-Verlag.

Arbib MA, editor. 1995. The handbook of brain theory and neural networks. London: MIT Press.

Arshavsky YI, Beloozerova IN, Orlovsky GN, Pavlova GA, Panchin YV. 1984. Neurons of pedal ganglia of pteropodial mollusc controlling activity of the locomotor generator. Neurophysiology. New York: Plenum. Eng. trans. from Neirofiziologiya (Kiev) 16:543–546.

Arshavsky YI, Beloozerova IN, Orlovsky GN, Panchin YV, Pavlova GA. 1985a. Control of locomotion in marine mollusc Clione limacina I. Efferent activity during actual and fictitious swimming. Exp Brain Res 58:225–262.

Arshavsky YI, Beloozerova IN, Orlovsky GN, Panchin YV, Pavlova GA. 1985b. Control of locomotion in marine mollusc Clione limacina II Rhythmic neurons of pedal ganglia. Exp Brain Res 58:263–272.

Arshavsky YI, Beloozerova IN, Orlovsky GN, Panchin YV, Pavlova GA. 1985c. Control of locomotion in marine mollusc Clione limacina III. On the ogin of locomotory rhythm. Exp Brain Res 58:273–284.

Arshavsky YI, Beloozerova IN, Orlovsky GN, Panchin YV, Pavlova GA. 1985d. Control of locomotion in marine mollusc Clione limacina IV. Role of type 12 interneurons. Exp Brain Res 58:285–292.

Arshavsky YI, Orlovsky GN, Panchin YV. 1985e. Control of locomotion in marine mollusc Clione limacina V. Photoinactivation of efferent neurons. Exp Brain Res 59:203–205.

Arshavsky YI, Gelfand IM, Orlovsky GN. 1986a. Cerebellum and rhythmic movements. Berlin: Springer-Verlag.

Arshavsky YI, Orlovsky GN, Panchin YV. 1986b. Control of locomotion in marine mollusc Clione limacina VI. Activity of isolated neurones of pedal ganglia. Exp Brain Res 63:106–112.

Ashby WE. 1952. Design for a brain. New York: Wiley.

Atwood HL, Wiersma CAG. 1967. Command interneurons in the crayfish central nervous system. J Exp Biol 46:249–261.

Baev KV. 1979. Depolarization of different lumbar afferent terminals during fictitious scratching. Neurophysiology. New York: Plenum. Eng. trans. from Neirofiziologiya (Kiev) 11:569–577.

Baev KV. 1980. Polarization of primary affarent terminals during fictitious locomotion. Neurophysiology. New York: Plenum. Eng. trans. from Neirofiziologiya (Kiev) 12:481–489.

Baev KV. 1981a. Reorganization of segmental responses to peripheral stimulation during fictitious scratching in cats. Neurophysiology. New York: Plenum. Eng. trans. from Neirofiziologiya (Kiev) 13:196–203.

Baev KV. 1981b. Retuning of segmental responses to peripheral stimulation in cats during fictitious locomotion. Neurophysiology. New York: Plenum. Eng. trans. from Neirofiziologiya (Kiev) 13:283–291.

Baev KV. 1991. Neurobiology of locomotion. Moscow: Nauka (in Russian).

Baev KV. 1994. Learning in systems controlling motor automatisms. Rev Neurosci 5:55–87.

Baev KV. 1995. Disturbances of learning processes in the basal ganglia in the pathogenesis of Parkinson's disease: a novel theory. Neurol Res 17:38–48.

Baev KV. 1997. Highest level automatisms in the nervous system: a theory of functional principles underlying the highest forms of brain function. Progr Neurobiol 51:129–166.

Baev KV, Beresovskii VK, Kebkalo TG, Savoskina LA. 1988. Afferent and efferent connections of brainstem locomotor regions. Neurosci 26:871–891.

Baev KV, Chub NI. 1989. The role of different spinal cord regions in generation of spontaneous motility in chick embryo. Neurophysiology. New York: Plenum. Eng. trans. from Neirofiziologiya (Kiev) 21:124–126.

Baev KV, Degtyarenko AM, Zavadskaya TV, Kostyuk PG. 1979. Activity of lumbar interneurons during late long-lasting discharges in motor nerves of immobilized thalamic cats. Neurophysiology. New York: Plenum. Eng. trans. from Neirofiziologiya (Kiev) 11:236–244.

Baev KV, Degtyarenko AM, Zavadskaya TV, Kostyuk PG. 1980. Activity of lumbar interneurons during fictitious locomotion in thalamic cats. Neurophysiology. New York: Plenum. Eng. trans. from Neirofiziologiya (Kiev) 11:329–338.

Baev KV, Degtyarenko AM, Zavadskaya TV, Kostyuk PG. 1981. Activity of lumbosacral interneurons during fictitious scratching. Neurophysiology. New York: Plenum. Eng. trans. from Neirofiziologiya (Kiev) 13:57–66.

Baev KV, Esipenko VB. 1988a. Intensity changes of integral afferent inflow from limb receptors and changes of primary afferent terminals polarization in a decerebrated cat during scratching. Neurophysiology. New York: Plenum. Eng. trans. from Neirofiziologiya (Kiev) 20:49–57.

Baev KV, Esipenko VB. 1988b. Limb movement parameters, intensity of the integral afferent impulsation from the limb receptors and level of primary afferent terminals polarization during locomotion in highly decerebrate cats. Neurophysiology. New York: Plenum. Eng. trans. from Neirofiziologiya (Kiev) 20:119–127.

Baev KV, Esipenko VB, Shimansky YP. 1991a. Afferent control of central pattern generators: experimental analysis of scratching in the decerebrate cat. Neurosci 40:239–256.

Baev KV, Esipenko VB, Shimansky YP. 1991b. Afferent control of central pattern generators: experimental analysis of locomotion in the decerebrate cat. Neurosci 43:237–247.

Baev KV, Kostyuk PG. 1981. Primary afferent depolarization evoked by activity of spinal scratching generator. Neurosci 6:205–215.

Baev KV, Kostyuk PG. 1982. Polarization of primary afferent terminals of lumbosacral cord elicited by activity of spinal locomotor generator. Neurosci 7:1401–1409.

Baev KV, Shimansky YP. 1992. Principles of organization of neural systems controlling automatic movements in animals. Progr Neurobiol 39:45–112.

Baev KV, Zavadskaya TV. 1981. Influence of bicuculline application on the upper cervical segments of the spinal cord and the scratch reflex. Physiol Journ 27:147–54 (in Russian).

Barto AG, Sutton RS, Anderson CW. 1983. Neuronlike elements that can solve difficult learning control problems. IEEE Trans Syst Man Cyber 13:835–846.

Batini C, Billard JM, Daniel H. 1985. Long-term modification of cerebellar inhibition after inferior olive degeneration. Exp Brain Res 59:404–409.

Bellman RE, Glicksberg I, Gross OA. 1958. Some aspects of the mathematical theory of control processes. Santa Monica, California: Rand Corporation.

Benabid AL, Pollack P, Gervason C, Hoffmann D, Gao DM, Hommel M, Perret J, de Rougemont J. 1991. Long-term suppression of tremor by chronic stimulation of the ventral intermediate thalamic nucleus. Lancet 337:403–406.

Beresovskii VK, Baev KV. 1988. New locomotor regions of the brainstem revealed by means of electrical stimulation. Neurosci 26:863–869.

Bergman H, Wichmann T, DeLong MR. 1990. Reversal of experimental parkinsonism by lesions of the subthalamic nucleus. Science 249:1436–1438.

Berkinblit MB, Deliagina TG, Feldman AG, Gelfand IM, Orlovsky GN. 1978. Generation of scratching I. Activity of spinal interneurons during scratching. J Neurophysiol 41:1040–1057.

Bernhard CG, Widen L. 1953. On the origin of the negative and positive spinal cord potentials evoked by stimulation of low threshold cutaneous fibres. Acta Physiol Scand 29 Suppl 106:42–54.

Bernstein NA. 1966. Sketches about physiology of movements and physiology of activity. Moscow: Meditzina (in Russian).

Bernstein NA. 1967. The co-ordination and regulation of movements. New York: Pergamon Press.

Bernstein NA. 1984. Human motor actions: Bernstein reassessed. New York: Elsevier.

Berridge MJ, Irvine RE. 1989. Inositol phosphates and cell signalling. Nature 341:197–205.

Bloedel JR, Lou J-S. 1987. The relation between Purkinje cell spike responses and the action of the climbing fibre system in unconditioned and conditioned responses of the forelimb to perturbed locomotion. In: Glickstein M, Yeo C, Stein J, editors. Cerebellum and neuronal plasticity. New York: Plenum. pp. 261–276.

Bowerman RF, Larimer JL. 1974a. Command fibers in the circumesophageal connectives of crayfish. 1. Tonic fibers. J Exp Biol 60:95–117.

Bowerman RF, Larimer JL. 1974b. Command fibers in the circumesophageal connectives of crayfish. 2. Phasic fibers. J Exp Biol 60:119–134.

Bracha V, Webster ML, Winters NK, Irwin KB, Bloedel JR. 1994. Effects of muscimol inactivation of the cerebellar interposed-dentate nuclear complex on the performance of the nictitating membrane response. Exp Brain Res 100:453–468.

Brodal A. 1981. Neurological anatomy. In relation to clinical medicine. New York: Oxford University Press.

Brodin L, Grillner S. 1985. The role of putative excitatory amino acid neurotransmitters in the initiation of locomotion in the lamprey spinal cord. 1. The effect of excitatory amino acid antagonists. Brain Res 360:139–148.

Brodin L, Grillner S, Rovainen C. 1985. NMDA, kinate and quisqualat receptors and the generation of fictive locomotion in the lamprey spinal cord. Brain Res 325:302–306.

Brown TG. 1914. On the nature of fundamental activity of the nervous centres; together with an analysis of the conditioning of rhythmic activity in progression, and a theory of evolution of function in the nervous system. J Physiol 48:18–46.

Brown TH, Kairiss EW, Keenan CL. 1990. Hebbian synapses: biophysical mechanisms and algorithms. Ann Rev Neurosci 13:475–511.

Buchanan JT. 1982. Identification of interneurons with collateral, caudal axons in the lamprey spinal cord: synaptic interactions and morphology. J Neurophysiol 47:961–975.

Buchanan JT. 1986. Premotor interneurons in the lamprey spinal cord: morphology, synaptic interactions and activities during fictive swimming. In: Neurobiology of vertebrate locomotion, Grillner S, Stein PSG, Stuart DG, Fossberg H, Herman R, editors. London: MacMillan Press. 321–333.

Buchanan JT, Cohen AH. 1982. Activities of identified interneurons, motoneurons and muscle fibers during fictive swimming in the lamprey and effects of reticulospinal and dorsal cell stimulation. J Neurophysiol 47:948–960.

Budakova NN. 1971. Stepping movements evoked by rhythmic stimulation of the dorsal root in the mesencephalic cat. Physiol Journ USSR 57:1632–1640 (in Russian).

Burgess N, Recce M, O'Keefe J. 1995. Hippocampus: spatial models. In: Arbib M, editor. The handbook of brain theory and neural networks. Cambridge, Massachusetts: A Bradford Book, MIT Press. 468–472.

Buzsaki G. Penttonen M, Nadasdy Z, Bragin A. 1996. Pattern and inhibition dependent invasion of pyramidal cell dendrites by fast spikes in the hippocampus in vitro. Proc Nat Acad Sci USA 93:9921–9925.

Byrne JH. 1987. Cellular analysis of associative learning. Physiol Rev 67:92–94.

Carew TJ, Sahley CL. 1986. Invertebrate learning and memory: from behavior to molecules. Ann Rev Neurosci 9:435–87.

Cerbone A, Sadile AG. 1994. Behavioral habituation to spatial novelty: interference and noninterference studies. Neurosci and Biobehav Rev 18:497–518.

Chub NL. 1991. Effect of L-DOPA on the spontaneous activity generated by isolated spinal cord of 16–20 day old chick embryo. Neurophysiology, New York: Plenum. Eng. trans. from Neirofiziologiya (Kiev) 23:338–343.

Chub NL, Baev KV. 1991. The influence of N-methyl-D-aspartate on spontaneous activity generated by isolated spinal cord of 16–20-day old chick embryos. Neurophysiology, New York: Plenum. Eng. trans. from Neirofiziologiya (Kiev) 24:205–213.

Churchland PS, Sejnowski TJ. 1992. The computational brain. Cambridge, Massachusetts: A Bradford Book, MIT Press.

Clark GA, McCormick DA, Lavond DG, Thompson RF. 1984. Effects of lesions of cerebellar nuclei on conditioned behavioral and hippocampal neuronal responses. Brain Res 291:125–136.

Cohen AH, Wallen P. 1980. The neuronal correlate of locomotion in fish. "Fictive swimming" induced in an in vitro preparation of the lamprey spinal cord. Exp Brain Res 41:11–18.

Dale N, Roberts A. 1984. Excitatory amino acid receptors in Xenopus embryo spinal cord and their role in the activation of swimming. J Physiol 348:527–543.

Dale N, Roberts A. 1985. Dual-component amino-acid-mediated synaptic potentials: excitatory drive for swimming in xenopus embryos. J Physiol 363:35–59.

Darian-Smith C, Darian-Smith I, Burman K, Ratcliffe N. 1993. Ipsilateral cortical projections to areas 3a, 3b, and 4 in the macaque monkey. J Comp Neurol 335:200–213.

Davis M. 1984. The mammalian startle response. In: Eaton RC, editor. Neural mechanisms in startle behavior. New York: Plenum Press. 287–351.

Davis J, Kennedy D. 1972a. Command interneurons controlling swimmeret movements in the lobster. 1. Types of effects on motor neurons. J Neurophysiol 35:1–12.

Davis J, Kennedy D. 1972b. Command interneurons controlling swimmeret movements in the lobster. 2. Interactions of effects on motor neurons. J Neurophysiol 35:13–19.

Davis J, Kennedy D. 1972c. Command interneurons controlling swimmeret movements in the lobster. 3. Temporal relationships among bursts in different motor neurons. J Neurophysiol 35:20–29.

Degtyarenko AM, Baev KV, Zavadskaya TV. 1990. Rearrangement of the efferent activity of a scratching generator under electrical activation of descending systems. Neurophysiology, New York: Plenum. Eng. trans. from Neirofiziologiya (Kiev) 22:300–309.

Degtyarenko AM, Zavadskaya TV. 1991. Rearrangement of the efferent activity of locomotor generator under electrical activation of descending systems in immobilized cats. Neurophysiology, New York: Plenum. Eng. trans. from Neirofiziologiya (Kiev) 23:151–160.

Degtyarenko AM, Zavadskaya TV, Baev KV. 1992. Mechanisms of supraspinal correction of scratching generator. Neurosci 46:189–195.

DeLong MR, Georgopoulos AP. 1981. Motor functions of the basal ganglia. In: Brookhart JM, Mountcastle VB, Brooks VB, editors. Handb. Physiol. Sect 1, The Nervous System. Vol 2, Motor Control Part 2. Bethesda: Am Physiol Soc. 1017–1061.

DeLong MR, Georgopoulos AP, Crutcher MD. 1983. Cortico-basal ganglia relations and coding of motor performance. Exp Brain Res, Suppl. 7:30–40.

Domer FR, Feldberg W. 1960. Scratching movements and facilitation of the scratch reflex produced by tubocurarine in cats. J Physiol 153:35–51.

Duenas SH, Rudomin P. 1988. Excitability changes of ankle extensor group Ia and Ib fibers during fictive locomotion in the cat. Exp Brain Res 70:15–25.

Dunsmore R, Lennox R. 1950. Stimulation and strychninization of supracallosal anterior cingulate gyrus. J Neurophysiol 13:207–214.

Eaton RC, Hacket JT. 1984. The role of the Mauthner cell in fast starts involving escape in teleost fish. In: Eaton RC, editor. Neural mechanisms of startle behavior. New York, London: Plenum 213–266.

Eccles J, Ito M, Szentagothai J. 1967. The cerebellum as a neuronal machine. New York: Springer-Verlag.

Edelman GM, Gall WE, Cowan WM, editors. (1987). Synaptic function. New York: Wiley.

Edgerton VR, Grillner S, Sjostrom A, Zangger P. 1975. The spinal generator for locomotion in the cat. Exp Brain Res Suppl 23:64.

Edgerton VR, Grillner S, Sjostrom A, Zangger P. 1976. Central generation of locomotion in vertebrates. In: Herman RM, Grillner S, Stein PSG, Stuart DG, editors. Neural control of locomotion. New York Plenum Press. 439–464.

Eidelberg E, Walden JG, Nguyen LH. 1981. Locomotor control in macaque monkeys. Brain 104:647–663.

Esipenko VB. (1987). Correlation between the kinematics of hindlimb movement and efferent activity in the decerebrate cat during scratching. Neurophysiology, New York: Plenum. Eng. trans. from Neirofiziologiya (Kiev) 19:525–533.

Evarts EV, Thach WT. 1969. Motor mechanism of the CNS: cerebrocerebellar inter-relations. Ann Rev Physiol 31:451–498.

Faber DS, Korn H. 1978. Electrophysiology of the Mauthner cell: basic properties, synaptic mechanisms, and associated networks. In: Faber DS, Korn H, editors. Neurobiology of the Mauthner cell New York: Raven Press. 47–131.

Feldberg W, Fleischauer K. 1960. Scratching movements evoked by drugs applied to the upper cervical cord. J Physiol 151:502–507.

Feldman AG. 1979. Central and reflex mechanisms of movement control. Moscow: Nauka (in Russian).

Feldman AG, Orlovsky GN, Perret C. 1977. Activity of muscle spindle afferents during scratching in the cat. Brain Res 129:192–196.

Felten DL, Sladek JR Jr. 1983. Monoamine distribution in primate brain V. Monoaminergic nuclei: anatomy, pathways and local organization. Brain Res Bull 10:171–284.

Ferraro G, Vella N, Sardo P, Caravaglios G, Sabatino M, La Grutta V. 1991. Dopaminergic control of feline hippocampal epilepsy: a nigro hippocampal pathway. Neurosci Lett 123:41–44.

Foy MR, Thompson RF. 1986. Single unit analysis of Purkinje cell discharge in classically conditioned and untrained rabbits. Soc Neurosci Abstr 12:518.

French JD, Hernandez-Peon R, Livingston RB. 1955. Projections from cortex to cephalic brain stem (reticular formation) in monkey. J Neurophysiol 18:74–95.

Garcia-Rill E, Skinner RD. 1986. The basal ganglia and the mesencephalic locomotor region. In: Grillner S, Stein PSG, Stuart DG, Fossberg H, Herman R, editors. Neurobiology of vertebrate locomotion. London: MacMillan Press. 77–103.

Garcia-Rill E, Skinner RD, Fitzgerald JA. 1985. Chemical activation of the mesencephalic locomotor region. Brain Res 330:43–54.

Garcia-Rill E, Skinner RD, Gilmore SA, Owings R. 1983. Connections of the mesencephalic locomotor region (MLR). 2. Afferents and efferents. Brain Res Bull 10:63–71.

Getting PA. 1981. Mechanisms of pattern generation underlying swimming in Tritonia. I. Neuronal network formed by monosynaptic connections. J Neurophysiol 46:65–79.

Getting PA. 1983a. Mechanisms of pattern generation underlying swimming in Tritonia. II. Network reconstruction. J Neurophysiol 49:1017–1035.

Getting PA. 1983b. Mechanisms of pattern generation underlying swimming in Tritonia. III. Intrinsic and synaptic mechanisms for delayed excitation. J Neurophysiol 49:1036–1050.

Getting PA. 1986. Understanding central pattern generators: insights gained from the study of invertebrate systems. In: Grillner S, Stein PSG, Stuart DG, Fossberg H, Herman R, editors. Neurobiology of vertebrate locomotion. London: MacMillan Press. 231–244.

Getting PA. 1989. Reconstruction of small neuronal networks. In: Koch C, Segev I, editors. Methods in neuronal modeling. Cambridge, Mass: MIT Press.

Getting PA, Dekin MS. 1985. Mechanisms of pattern generation underlying swimming in Tritonia. IV. Gating of a central pattern generator. J Neurophysiol 53:466–480.

Getting PA, Lennard PR, Hume RI. 1980. Central pattern generator mediating swimming in Tritonia. I. Identification and synaptic interactions. J Neurophysiol 44:151–164.

Gibson AR, Robinson FR, Alam J, Houk JC. 1987. Somatotopic alignment between climbing fiber input and nuclear output of the cat intermediate cerebellum. J Comp Neurol 260:362–377.

Granit R. 1970. The basis of motor control. New York: Academic Press.

Greene KA, Marciano F, Golfinos JG, Shetter AG, Lieberman AN, Spetzler RF. 1992. Pallidotomy in levodopa era. Adv Clinical Neurosci 2:257–281.

Grillner S. 1974. On the generation of locomotion in the spinal dogfish. Exp Brain Res 20:159–170.

Grillner S. 1975. Locomotion in vertebrates: central mechanisms and reflex interaction. Physiol Rev 55:247–304.

Grillner S. 1976. Some aspects of the descending control of the spinal circuits generating locomotor movements. In: Herman RM, Grillner S, Stein PSG, Stuart DG, editors. Neural control of locomotion. New York: Plenum Press. 351–375.

Grillner S. 1996. Neural networks for vertebrate locomotion. Sci Amer 274:64–69.

Grillner S, Brodin L, Sigvardt K, Dale N. 1986. On the spinal network generating locomotion in lamprey: transmitters, membrane properties and circuitry. In: Grillner S, Stein PSG, Stuart DG, Fossberg H, Herman R, editors. Neurobiology of vertebrate locomotion. London: MacMillan Press. 335–352.

Grillner S, Deliagina T, Ekeberg O, El Manira A, Lansner A, Orlovsky GN, Wallen P. 1995. Neural networks that co-ordinate locomotion and body orientation in lamprey. TINS 18:270–279.

Grillner S, Kashin S. 1976. On the generation and performance of swimming in fish. In: Herman RM, Grillner S, Stein PSG, Stuart DG, editors. Neural control of locomotion. New York: Plenum Press. 181–202.

Grillner S, McClellan A, Sigvardt K. 1982. Mechanosensitive neurons in the spinal cord of the lamprey. Brain Res 235:169–173.

Grillner S, Wallen P. 1985. Central pattern generators for locomotion, with special reference to vertebrates. Ann Rev Neurosci 8:233–261.

Grillner S, Zangger P. 1974. Locomotor movements generated by the deafferented spinal cord. Acta Physiol Scand 91:38A–39A.

Grossberg S, Kuperstein M. 1989. Neural dynamics of adaptive sensory-motor control. New York: Pegamon Press.

Grossberg S, Merrill JWL. 1996. The hippocampus and cerebellum in adaptively timed learning, recognition, and movement. J Cogn Neurosci 8:257–277.

Gurfinkel VS, Shik ML. 1973. The control of posture and locomotion. In: Gydikov AA et al, editors. Motor control. New York: Plenum 217–234.

Hamburger V. 1963. Some aspects of the embryology of behavior. Quart Rev Biol 38:342–365.

Hamburger V. 1975. Cell death in the development of the lateral motor column of the chick embryo. J Comp Neurod 160:535–546.

Hamburger V, Oppenheim R. 1967. Prehatching motility and hatching behavior in the chick. J Exp Zool 166:171–204.

He SQ, Dum RP, Strick PL. 1995. Topographic organization of corticospinal projections from the frontal lobe: motor areas on the medial surface of the hemisphere. J Neurosci 15:3284–3306.

Hebb DO. 1949. The organization of behavior. New York: Wiley.

Hecht-Nielsen R. 1990. Neurocomputing. San Diego: Addison-Wesley Publishing Company.

Heimer L. 1978. The olfactory cortex and the ventral striatum. In Livingston KE, Hornykiewicz O, editors. Limbic mechanisms. New York: Plenum 95–187.

Heimer L, Wilson RD. 1975. The subcortical projections of the allocortex. Similarities in the neural associations of the hippocampus, the piriform cortex, and the neocortex. In: Santini M, editor. Golgi Centennial Symposium: Perspectives in neurology. New York: Raven 177–193.

Hemphill M, Holm G, Crutcher M, DeLong MR, Hedreen J. 1981. Afferent connections of the nucleus accumbens in the monkey. In: Chronister R, DeFrance J, editors. Neurobiology of the nucleus accumbens. Brunswick, Maine: Haer Inst. Press 75–81.

Hoover JE, Strick PL. 1993. Multiple output channels in the basal ganglia. Science 259:819–821.

Hopfield JJ. 1984. Neurons with graded response have collective computational properties like those of two-state neurons. Proc. of the National Acad. of Sci. USA 81:3088–3092.

Hopfield JJ. 1987. Learning algorithms and probability distributions in feed-forward and feed-back networks. Proc. of the National Acad. of Sci. USA 84:8429–8433.

Hopfield JJ. 1994. Neurons, dynamics and computation. Physics Today 47:40–46.

Hopfield JJ, Tank DW. 1985. "Neural" computation of decisions in optimization problems. Biol Cybern 52:141–152.

Houk JC, Adams JL, Barto AG. 1995. A model of how the basal ganglia generate and use neural signals that predict reinforcement. In: Houk JC, Davis JL, Beiser DG, editors. Models of information processing in the basal ganglia. Cambridge, Massachusetts: Bradford Book, MIT Press 249–270.

Hounsgaard J, Hultborn H, Jespersen B, Kiehn O. 1984. Intrinsic membrane properties causing a bistable behaviour of a-motoneurons. Exp Brain Res 55:391–394.

Hounsgaard J, Kiehn O. 1985. Ca dependent bistability induced by serotonin in spinal motoneurons. Exp Brain Res 57:422–425.

Hoyle G. 1984. The scope of neuroethology. Behav and Brain Sci 7:367–412.

Huang CC, Lan CM, Peng MT. 1970. Elicitation of scratching movements by mechanical stimulation to the spinal cord in decerebrate cats. J Physiol 20:365–369.

Hughes GM, Wiersma CAG. 1960. The coordination of swimmeret movements by the crayfish, Procambarus Clarkii (Girard). J Exp Biol 37:657–670.

Isaacson RL. 1982. The limbic system. New York: Plenum Press.

Jankowska E, Jukes MGM, Lund S, Lundberg A. 1967. The effect of DOPA on the spinal cord. 6. Half-centre organization of interneurones transmitting effects from the flexor reflex afferents. Acta Physiol Scand 70:389–402.

Johnson JW, Ascher P. 1987. Glycine potentiates the NMDA response in cultured mouse brain neurons. Nature 325:529–531.

Jordan LM. 1986. Initiation of locomotion from the mammalian brainstem. In: Grillner S, Stein PSG, Stuart DG, Fossberg H, Herman R, editors. Neurobiology of vertebrate locomotion. London: MacMillan Press 21–37.

Kalman R, Bucy R. 1961. New results in linear filtering and prediction theory. J Basic Engr (ASME Trans) 83:95–108.

Kandel ER, Hawkins RD. 1992. The biological basis of learning and individuality. Sci Amer 267:78–82.

Kandal ER, Schwartz JH, Jessell TM, editors. 1991. Principles of neural science. New York: Elsevier.

Kashin S, Brill R, Ikehara W, Dizon A. 1981. Induced locomotion by midbrain stimulation in restrained Skipjack tuna, Katwuwonus pelamis. J Exp Zool 216:327–329.

Kashin S, Feldman AG, Orlovsky GN. 1975. Locomotion of fish evoked by electrical stimulation of the brain. Brain Res 82:41–47.

Kawato M, Gomi H. 1992. The cerebellum and VOR/OKR learning models. TINS 15:445–453.

Kazennikov OV, Selionov VA, Shik ML, Yakovleva GV. 1980. Rhombencephalic "locomotor region" in turtle. Neurophysiology, New York: Plenum. Eng. trans. from Neirofiziologiya (Kiev) 12:328–330.

Kazennikov OV, Shik ML, Yakovleva GV. 1980. On the two ways for "locomotor influence" of the brainstem on the spinal cord. Physiol Journ USSR 66:1260–1263.

Kehoe JS. 1972a. Ionic mechanisms of a two-component cholinergic inhibition in Aplysia neurones. J Physiol 225:85–114.

Kehoe JS. 1972b. The physiological role of three acetylcholine receptors in synaptic transmission in Aplysia. J Physiol 225:147–172.

Kelley AE, Domesick VB. 1982. The distribution of the projection from the hippocampal formation to the nucleus accumbens in the rat: an anterograde- and retrograde-horseradish peroxidase study. Neurosci 7:2321–2335.

Kelley AE, Domesick VB, Nauta WJH. 1982. The amygdalostriatal projection in the rat—an anatomical study by anterograde and retrograde tracing methods. Neurosci 7:615–630.

Kelly TM, Zuo C-C, Bloedel JR. 1990. Classical conditioning of the eyeblink reflex in the decerebrate-decerebellate rabbit. Behav Brain Res 38:7–18.

Kemp JM, Powell TPS. 1971. The connections of the striatum and globus pallidus: synthesis and speculation. Philos Trans R Soc, London, Ser, B, 262:441–457.

Kling U. 1971. Stimulation neuronaler impulsrhythmen. Zur Theorie der Netzwerke mit cyclischen Hemmverbindungen. Kybernetic 9:123–139.

Kolmogorov AN. 1957. On the representation of continuous functions of many variables by superposition of continuous functions of one variable and addition. Dokl. Akad. Nauk USSR 114:953–956 (in Russian).

Kostyukov AI. 1987. Muscle dynamics: dependence of muscle length on changes in external load. Biol Cybern 56:375–387.

Kostyukov AI, Cherkassky VL. 1992. Movement-dependent after-effects in the firing of the spindle endings from the de-efferented muscles of the cat hindlimb. Neurosci 46:989–999.

Krayniak PF, Meiback R, Siegel A. 1981. A projection from the entorhinal cortex to the nucleus accumbens in the rat. Brain Res 209:427–431.

Kudo N, Yamada T. 1987. N-Methyl d, 1-aspartate-induced locomotor activity in a spinal cord hindlimb muscles preparation of the newborn rat studied in vitro. Neurosci Lett 75:43–48.

Kulberg AY, Ivanovska ND, Tarkhanova IA. 1987. Isolated extracellular domain of receptor for protein produced by a certain cell behaves as a specific inhibitor of protein biosynthesis. Immunology N1:25–27 (in Russian).

Kupferman I, Weiss K. 1978. The command neuron concept. Behav Brain Sci 1:3–39.

Kurkova V. 1995. Kolmogorov's theorem. In: Arbib M, editor. The handbook of brain theory and neural networks. Cambridge, Massachusetts: A Bradford Book, MIT Press 501–502.

Laitinen LV. 1985. Brain targets in surgery for Parkinsons disease. J Neurosurg 62:349–351.

Laitinen LV, Bergenheim AT, Hariz MI. 1992. Leksells posteroventral pallidotomy in the treatment of Parkinsons disease. J Neurosurg 76:53–61.

Lavond DG, McCormick DA, Clark GA, Holmes DT, Thompson RF. 1981. Effects of ipsilateral rostral pontine reticular lesions on retention of classically conditioned nictitating membrane and eyelid response. Physiol Psychol 9:335–339.

Lavond DG, Hembree TL, Thompson RF. 1985. Effect of kainic acid lesions of the cerebellar interpositus nucleus on eyelid conditioning in the rabbit. Brain Res 326:179–182.

Leigh JR. 1987. Applied control theory. London: Peregrinus on behalf of the Institution of Electrical Engineers.

Le Moal M, Simon H. 1991. Mesocorticolimbic dopaminergic network: functional and regulatory roles. Physiol Rev 71:155–234.

Leonard RB, Rudomin P, Droge MH, Grossman AE, Willis WD. 1979. Locomotion in the decerebrate stingray. Neurosci Lett 14:315–319.

Levy W. 1989. A computational approach to hippocampal function. In: Hawkins RD, Bower GH, editors Computational Models of Learning in Simple Neural Systems, San Diego: Academic Press 243–30 5.

Levy WB, Wu XB, Baxter RA. 1995. A computational model of hippocampal region CA3 for flexible learning of temporal spatial associations. Soc Neurosci Abstr 21:1226.

Lidierth M, Apps R. 1990. Gating in the spino-olivo-cerebellar pathways to the c zone of the cerebellar cortex during locomotion in the cat. J Physiol 430:453–469.

Lincoln JS, McCormick DA, Thompson RF. 1982. Ipsilateral cerebellar lesions prevent learning of the classically conditioned nictitating membrane/eyelid response of the rabbit. Brain Res 242:190–193.

Lindblom UF, Ottosson JO. 1953. Localization of the structure generating the negative cord dorsum potential evoked by stimulation of low threshold cutaneous fibres. Acta Physiol Scand. Suppl 106:180–190.

Lissman HW. 1946. The neurological basis of the locomotory rhythm in the spinal dogfish. J Exp Biol 23:162–176.

Ljungberg T, Apicella P, Schultz W. 1992. Responses of monkey dopamine neurons during learning of behavioral reactions. J Neurophysiol 67:145–163.

Lorenz K. 1950. The comparative method in studying innate behavior patterns. Symp Soc Exp Biol 4:221–268.

Lou J-S, Bloedel JR. 1988. A new conditioning paradigm: conditioned limb movements in locomoting decerebrate ferrets. Neurosci Lett 84:185–190.

Lu MT, Preston JB, Strick PL. 1994. Interconnections between the prefrontal cortex and the premotor areas in the frontal lobe. J Comp Neurol 341:375–392.

Magnus R. 1989. Body posture: experimental-physiological investigations of the reflexes involved in body posture, their cooperation and disturbances. New York: Amerind Publishing Company.

Marciano FF, Greene KA. 1992. Surgical management of Parkinsons disease. Part I: Paraneural and neural tissue transplantation. Neurol Forum 3:1–7.

Marr D. 1969. A theory of cerebellar cortex. J Physiol. 202:437–470.

Mauk MD, Thompson RF. 1987. Retention of classically conditioned eyelid responses following acute decerebration. Brain Res 403:89–95.

Maxwell DJ, Koerber HR. 1986. Fine structure of collateral axons originating from feline spinocervical tract neurons. Brain Res 363:199–203.

McClellan AD. 1986. Command systems for initiating locomotion in fish and amphibians: parallels to initiation systems in mammals. In: Grillner S, Stein PSG, Stuart DG, Fossberg H, Herman R, editors Neurobiology of vertebrate locomotion. London: MacMillan Press 3–20.

McClellan AD, Farel PB. 1985. Pharmacological activation of locomotor patterns in larval and adult frog spinal cords. Brain Res 332:119–130.

McClellan AD, Grillner S. 1983. Initiating and sensory gating of "fictive" swimming and withdrawal responses in an in vitro preparation of the lamprey spinal cord. Brain Res 269:237–250.

McClellan AD, Grillner S. 1984. Activation of "fictive" swimming by electrical microstimulation of "locomotor command regions" in the brainstem of the lamprey. Brain Res 300:352–362.

McCormick DA, Clark GA, Lavond DG, Thompson RF. 1982. Initial localization of the memory trace for a basic form of learning. Proc Natl Acad Sci USA 79:2731–2742.

McCormick DA, Lavond DG, Clark GA, Kettner RE, Rising CE, Thompson RF. 1981. The engram found? Role of the cerebellum in classical conditioning of nictitating membrane and eyelid responses. Bull Psychon Soc 18:103–105.

McCormick DA, Steinmetz JE, Thompson RF. 1985. Lesions of the inferior olivary complex cause extinction of the classically conditioned eyeblink response. Brain Res 359:120–130.

McCulloch W, Pitts W. 1943. A logical calculus of the ideas immanent in nervous activity. Bull Math Biophys 5:115–137.

Merton PA. 1953. Speculations on the servo-control of movement. In: Wolstenholme GEW, editor. The Spinal cord. London: Churchill 247–260.

Mesarovic MD, Macko D, Takahara Y. 1970. Theory of hierarchical multilevel systems. New York: Academic Press.

Miller S, Scott PD. 1977. The spinal locomotor generator. Exp Brain Res 30:387–403.

Minsky ML. 1963. Steps towards artificial intelligence. In: Feigenbaum EA, Feldman J, editors. Computers and thought. New York: McGraw-Hill 406–450.

Moffett J, Kratz E, Florkiewicz R, Stachowiak MK. 1996. Promoter regions involved in density-dependent regulation of basic fibroblast growth factor gene expression in human astrocytic cells. Proc Natl Acad Sci USA 93:2470–2475.

Montgomery EB, Gorman DS, Nuessen J. 1991. Motor initiation versus execution in normal and Parkinson's disease subjects. Neurology 41:1469–1475.

Mori S, Shik ML, Yagodnitsyn AS. 1977. Role of pontine tegmentum for locomotor control in mesencephalic cat. J Neurophysiol 40:284–295.

Nauta WJH. 1961. Fibre degeneration following lesions of the amygdaloid complex in the monkey. J Anat 95:515–531.

Nauta HJW. 1979. A proposed conceptual reorganization of the basal ganglia and telencephalon. Neurosci 4:1875–1881.

Nauta WJH, Domesick VB. 1984. Afferent and efferent relationships of the basal ganglia. In: Functions of the basal ganglia. Ciba Found Symp 107:3–29.

Noga BR, Kettler J, Jordan LM. 1988. Locomotion produced in mesencephalic cat by injection of putative transmitter substances and antagonists into the medial reticular formation and the pontomedullary locomotor strip. J Neurosci 8:2074–2086.

Norman RJ, Buchwald JS, Villablanca JR. 1977. Classical conditioning with auditory discrimination of the eyeblink in decerebrate cats. Science 196:551–553.

Oakley DA, Russell IS. 1972. Neocortical lesions and classical conditioning. Physiol Behav 8:915–926.

O'Donovan MJ. 1987. Developmental approach to the analysis of vertebrate central pattern generators. J Neurosci Meth 21:275–286.

Oman CM. 1988. Motion sickness: A synthesis and evaluation of the sensory conflict theory. Can. J. Physiol. Pharmacol. 68: 294–303.

Orlovsky GN. 1970. On the connections of reticulo-spinal neurons with "locomotor regions". Biophysics 15:171–178 (in Russian).

Orlovsky GN, Feldman AG. 1972. Classification of lumbosacral neurons according to their discharge patterns during evoked locomotion. Neurophysiology, New York: Plenum. Eng. trans. from Neirofiziologiya (Kiev) 4:410–417.

Ostriker G, Pellionisz A, Llinas R. 1982. Tensor network theory applied to the oculomotor system: CNS activity expressed with natural, non-orthogonal coordinates. Soc Neurosci Abstr 12:N 45.2.

Panchin YuV, Skrima RN. 1978. Scratch reflex evoked by application of strychnine on the spinal cord. Neurophysiology, New York: Plenum. Eng. trans. from Neirofiziologiya (Kiev) 10:622–625.

Pavlov IP. 1949. Selected works. Moscow: Governmental Publishing House of Political Literature: (in Russian).

Perret C. 1976. Neural control of locomotion in the decorticate cat. In. Herman RM, Grillner S, Stein PSG, Stuart DG, editors. Neural control of locomotion. New York: Plenum Press 587–615.

Philippson M. 1905. L'autonomie et la centralization dans le system nerveux des animaux. Trav Lab Physiol Inst Solvay, Bruxelles 7:1–208.

Pittman R, Oppenheim RW. 1979. Cell death of motoneurons in the chick embryo spinal cord. IV. Evidence that a functional neuromuscular interaction is involved in the regulation of naturally occurring cell death and the stabilization by synapses. J Comp Neurol 187:425–446.

Pontryagin LS. 1990. Optimal control and differential games: collection of papers. Providence, R.I.: American Mathematical Society.

Poulos CX, Sheafor PJ, Gormezano I. 1971. Classical appetitive conditioning of the rabbit's (Oryctolagus Cuniculus) jaw-movement response with a single-alternation schedule. J Comp Psychol 75:231–238.

Pribram KH, MacLean PD. 1953. Neuronographic analysis of medial and basal cortex: II. Monkey. J Neurophysiol 16:323–340.

Ripley KL, Provine RR. 1972. Neural correlates of embryonic motility in the chick. Brain Res 45:127–134.

Roberts A, Kahn JA, Soffe SR, Clarke JDW. 1981. Neural control of swimming in a vertebrate. Science 213:1032–1034.

Roberts A, Soffe SR, Dale N. 1986. Spinal interneurones and swimming in frog embryos. In: Grillner S, Stein PSG, Stuart DG, Fossberg H, Herman R, editors. Neurobiology of vertebrate locomotion. London: MacMillan Press 279–306.

Roberts A, Tunstall J. 1990. Mutual re-excitation with post-inhibitory rebound: a simulation study on the mechanisms for locomotor rhythm generation in the spinal cord of Xenopus embryos. Eur J Neurosci 2:11–23.

Roberts A, Tunstall MJ, Wolf E. 1995. Properties of networks controlling locomotion and significance of voltage dependency of NMDA channels: simulation study of rhythm generation sustained by positive feedback. J Neurophysiol 73:485–495.

Rosenblatt F. 1962. Principles of neurodynamics; perceptrons and the theory of brain mechanisms. Washington: Spartan Books.

Rovainen CM. 1974 Synaptic interactions of identified nerve cells in the spinal cord of the sea lamprey. J Comp Neurol 154:189–206.

Rovainen CM. 1979. Neurobiology of lamprey. Physiol Rev 59:1007–1077.

Rovainen CM. 1986. The contributions of multisegmental interneurons to the longitudinal coordination of fictive swimming in the lamprey. In: Grillner S, Stein PSG, Stuart DG, Fossberg H, Herman R, editor Neurobiology of vertebrate locomotion. London: MacMillan Press 353–370.

Rusin KI, Baev KV. 1990. Effect of noradrenaline precursors on glycine and NMDA receptors of chick embryo spinal cord neurons. Neurophysiology, New York: Plenum. Engl. trans. from Neirofiziologiya (Kiev) 22:665–670.

Rusin KI, Baev KV, Batueva IV, Safronov BV, Suderevskaya EI. 1989. Action of L-DOPA on spinal cord neurons in lamprey Lampetra fluviatilis. J Evol Biochem and Physiol 25:404–405 (in Russian).

Schmajuk NA. 1995. Cognitive maps. In: Arbib M, editor. The handbook of brain theory and neural network. Cambridge, Massachusetts: A Bradford Book, MIT Press 197–200.

Schmajuk NA, Thieme AD. 1992. Purposive behavior and cognitive mapping: an adaptive neural network. Biol Cybern 67:165–174.

Schmajuk NA, Thieme AD, Blair HT. 1993. Maps, routes, and the hippocampus: a neural network approach. Hippocampus 3:387–400.

Schmidt, RF. 1971. Presynaptic inhibition in the vertebrate central nervous system. Ergeb. Physiol. 63:20–101.

Schomburg ED. 1990. Spinal sensorimotor systems and their supraspinal control. Neurosci Res 7:265–340.

Schrameck NE. 1970. Crayfish swimming: alternating motor output and giant fiber activity. Science 169:698–700.

Sears LL, Steinmetz JE. 1991. Dorsal accessory inferior olive activity diminishes during acquisition of the rabbit classically conditioned eyelid response. Brain Res 545:114–122.

Sechenov IM. 1863. Brain reflexes. Moscow: Publishing House of the Academy of Medical Sciences, 1952 (in Russian).

Selverston AI. 1980. Are central pattern generators understandable? Behav Brain Sci 3:535–571.

Severin FV. 1970. The role of the gamma-motor system in the activation of the extensor alpha motor neurones during controlled locomotion. Biophysics, Eng. trans. from Biofizika (Moscow) 12:502–511.

Shannon CE. 1958. Von Neumann's contributions to automata theory. Bull Amer Math Soc 64:139–173.

Shepherd GM. 1994. Neurobiology. New York: Oxford University Press.

Sherrington CS. 1906. Observations on the scratch-reflex in the spinal dog. J Physiol 34:1–50.

Sherrington CS. 1910a. Flexion-reflex of the limb, crossed-extension reflex, and reflex stepping and standing. J Physiol 40:28–121.

Sherrington CS. 1910b. Notes on the scratch-reflex of the cat. Quart J Exp Physiol 3:213–220.

Sherrington CS. 1947. The integrative action of the nervous system. New Haven: Yale University Press.

Shik ML. 1976. Control of terrestrial locomotion in mammals. In: Physiology of movements. Leningrad: Nauka 234–275 (in Russian).

Shik ML, Orlovsky GN. 1976. Neurophysiology of locomotor automatism. Physiol Rev 56:465–501.

Shik ML, Orlovsky GN, Severin FV. 1966a. Organization of locomotor synergy. Biophysics 11:879–886 (in Russian).

Shik ML, Severin FV, Orlovsky GN. 1966b. Control of walking and running by electrical stimulation of the midbrain. Biophysics 11:659–666 (in Russian).

Shimansky YP. 1987. Reordering of scratch generator efferent activity produced by cyclic hindlimb movement in decerebrate immobilized cat. Neurophysiology, New York: Plenum. Eng. trans. from Neirofiziologiya (Kiev) 19:443–449.

Shimansky YP, Baev KV. 1986. Dependence of the efferent activity parameters on the hindlimb position during fictitious scratching in decerebrated cat. Neurophysiology, New York: Plenum. Eng. trans. from Neirofiziologiya (Kiev) 18:636–645.

Shimansky YP, Baev KV. 1987a Rebuilding of scratching generator efferent activity under the influence of phasic stimulation of hindlimb muscle afferents in decerebrated immobilized cat. Neurophysiology, New York: Plenum. Eng. trans. from Neirofiziologiya (Kiev) 19:372–382.

Shimansky YP, Baev KV. 1987b. Rebuilding of scratching generator efferent activity under the influence of phasic stimulation of hindlimb skin afferents in decerebrated immobilized cat. Neurophysiology, New York: Plenum. Eng. trans. from Neirofiziologiya (Kiev) 19:382–390.

Siegelbaum SA, Kandel ER. 1991. Learning-related synaptic plasticity. Curr Opin Neurobiol 1:113–120.

Siegfried J, Lipitz B. 1995. Bilateral chronic electrostimulation of ventroposterolateral pallidum: a new therapeutic approach for alleviating all parkinsonian symptoms. Neurosurg 35:1126–1130.

Sillar KT, Skorupski P. 1986. Central input to primary afferent neurons in crayfish, Pacifastacus leniusculus, is correlated with rhythmic motor output of thoracic ganglia. J Neurophysiol 55:678–688.

Sillar KT, Scorupski P, Elson RC, Bush BM. 1986. Two identified afferent neurones entrain a central locomotor rhythm generator. Nature 323:440–443.

Sjostrom A, Zangger P. 1976. Muscle spindle control during locomotor movements generated by the deafferented spinal cord. Acta Physiol Scand 97:281–291.

Skinner RD, Garcia-Rill E. 1985. The mesencephalic locomotor region (MLR) in the rat. Brain Res 323:385–389.

Smith DO. 1974. Central nervous control of excitatory and inhibitory neurons of opener muscle of the crayfish claw. J Neurophysiol 37:108–118.

Soffe SR. 1990. Active and passive membrane properties of spinal cord neurons that are rhythmically active during swimming in Xenopus embryos. Eur J Neurosci 2:1–10.

Soffe SR, Clarke JDW, Roberts A. 1984. Activity of comissural interneurones in the spinal cord of Xenopus embryos. J Neurophysiol 51:1257–1267.

Soffe SR, Roberts A. 1982a. Activity of myotomal motoneurones during fictive swimming in frog embryos. J Neurophysiol 48:1274–1278.

Soffe SR, Roberts A. 1982b. Tonic and phasic synaptic input to spinal cord motoneurones during fictive locomotion in frog embryos. J Neurophysiol 48:1279–1288.

Soffe SR, Roberts A. 1989. The influence of magnesium ions on the mediated responses of ventral rhythmic neurons in the spinal cord of Xenopus embryos. Eur J Neurosci 1:507–515.

Spector NH. 1987. Old and new strategies in the conditioning of immune responses. In: Neuroimmune interactions: proceedings of the second international workshop on neuroimmunomodulation. New York: New York Academy of Sciences 522–531.

Spiegel EA. 1966. Development of stereoencephalotomy for extrapyramidal diseases. J Neurosurg. Suppl 24:433–439.

Steeves JD, Jordan LM. 1980. Localization of descending pathway in the spinal cord which is necessary for controlled treadmill locomotion. Neurosci Lett 20:283–288.

Steeves JD, Weinstein GN. 1984. Brainstem areas and descending pathways for the initiation of flying and walking in birds. Soc Neurosci Abstr 10:30.

Stein PSG. 1971. Intersegmental coordination of swimmeret motor neuron activity in crayfish. J Neurophysiol 34:310–318.

Stein RB. 1982. What muscle variables does the nervous system control in limb movements? Behav Brain Sci 5:535–571.

Steinmetz JA, Lavond DG, Thompson RF. 1985. Classical conditioning of the rabbit eyelid response with mossy fiber stimulation as the conditioned stimulus. Bull Psychon Soc 23(3):245–248.

Swain RA, Thompson R. 1993. In search for engrams. Ann NY Acad Sci 17:27–39.

Szekely G. 1968. Development of limb movements: embryological, physiological and model studies. In: Wolstenholme GEW, Connor MO, editors. Ciba Found. Symp. Growth of the Nervous System. London: Churchill 77–93.

Tasker RR. 1990. Thalamotomy. Neurosurg Clin North Am 1:841–864.

Thompson RF. 1989. Role of the inferior olive in classical conditioning. Exp Brain Res 17:347–362.

Tolman EC. 1932. Cognitive maps in rats and men. Psychol Rev 55:189–208.

Tou JT, Gonzalez RC. 1974. Pattern recognition principles. Don Mills, Ontario: Addison-Wesley Publishing Company.

Tsubokawa H, Ross WN. 1996. IPSPs modulate spike backpropagation and associated $[Ca^{2++}]$ changes in the dendrites of hippocampal CA1 pyramidal neurons. J Neurophysiol 76:2896–2906.

Turing AM. 1936. On computable numbers with an application to the Entscheidungsproblem. Proc London Math Soc Ser. 2, 42:230–265, and also corrections: Proc London Math Soc. 1937. 43:544–545.

Turing AM. 1950. Computing machinery and intelligence. Mind 59:433–460.

Ungerleider LG. 1995. Functional brain imaging studies of cortical mechanisms for memory. Science 270:769–775.

Vankov A, Herve-Minvielle A, Sara SJ. 1995. Responses to novelty and its rapid habituation in locus coeruleus neurons of the freely exploring rat. Eur J Neurosci 7:1180–1187.

Viala D, Buisseret-Delmas C, Portal JJ. 1988. An attempt to localize lumbar locomotor generator in the rabbit using 2-deoxy [^{14}C] glucose autoradiography. Neurosci Lett 86:139–143.

Viala D, Valin A, Buser P. 1974. Relation between the "late reflex discharge" and locomotor movements in acute spinal cats and rabbits treated with DOPA. Arch Ital Biol 112:299–306.

Viala G, Viala D. 1977. Elements of locomotor programming in the rabbit. In: Proc. XXVII Int. congr. physiol. sci. satell. symp. "Neurophysiological mechanisms of locomotion," Paris.

von Bertalanffy L. 1950. An outline of general system theory. Brit J Philos Sci 1:134–164.

von Holst E. 1954. Relations between the central nervous system and the peripheral organs. Brit J Anim Behav 2:89–94.

von Holst E, Mittelstaedt H. 1950. Das Reafferenzprinzip: Wechselwiskungen zwischen Zentaralnervensystem und Peripherie. Naturwissenschaften 37:464–476.

von Neumann J, Morgenstern O. 1953. Theory of games and economic behavior. Princeton: Princeton University Press.

Weeks JC, Kristan WB. 1978. Initiation, maintenance and modulation of swimming in the medical leech by the activity of a single neuron. J Exp Biol 77:71–88.

Welker W, Blair C, Shambes GM. 1988. Somatosensory projections to cerebellar granule cell layer of giant bushbaby, Galago crassicaudatus. Brain Behav Evol 31:150–160.

Wiener N. 1961. Cybernetics; or, Control and communication in the animal and the machine. New York: MIT Press.

Wiersma CAG. 1938. Function of the giant fibers of the central nervous system of the crayfish. Proc Soc Exp Biol and Med 38:661–662.

Wiersma CAG, Ikeda K. 1964. Interneurons commanding swimmeret movements in the crayfish Procambarus Clarkii (Girard). Comp Biochem and Physiol 12:509–525.

Williams BJ, Livingston CA, Leonard RB. 1984. Spinal cord pathways involved in initiation of swimming in the stingray, Dasyatis sabina: spinal cord stimulation and lesions. J Neurophysiol 51:578–591.

Wine JJ, Krasne FB. 1972. The organization of escape behavior in the crayfish. J Exp Biol 56:1–18.

Wolpaw JR, Herchenroder PA. 1990. Operant conditioning of H-reflex in freely moving monkeys. J Neurosci Meth 31:145–152.

Wolpaw JR, Lee CL, Carp JS. 1991. Operantly conditioned plasticity in spinal cord. Ann NY Acad Sci 627:338–348.

Wolpert DM, Ghahramani Z, Jordan M. 1995. An internal model for sensorimotor integration. Science 269:1880–1882.

Yeo CH, Hardiman MJ, Glickstein M. 1985. Classical conditioning of the nictitating membrane response of the rabbit. I. Lesions of the cerebellar nuclei. Exp Brain Res 60:87–98.

Appendix

1 The Main Properties of Sensory Information Sources and Channels

Because the sensory signal passing through a single information channel (afferent fiber) carries only approximate data about the current phase state of the controlled object, it may be represented from the mathematical point of view as a random value distributed over the phase state space. Thus, a simple duplication of information in different channels indicates the identity of the mathematical expectations of the corresponding random values. The way to obtain the maximum reliable estimation in this case is well known from the theory of probability. If there are n random variables, which have equal mathematical expectation and dispersion, then the dispersion can be diminished $\sqrt{n}$ by simple averaging. In a more general case, when random variables $x_1, x_2, \ldots, x_n$ have the same mathematical expectation and different dispersions, it is easy to show that their linear combination (weighted sum)

$$\sum_{i=1}^{n} k_i x_i \left(\text{where } \sum_{i=1}^{n} k_i = 1 \text{ and } k_i \geq 0, i = 1, \quad , n \right)$$

which has the same mathematical expectation, will have minimum dispersion (among linear sums) if k_i is inversely proportional to the corresponding dispersion D_i. Having introduced "reliability" $r_i = 1/D_i$ we can write the simple expression for the maximum reliable weighted mean of the random variables:

$$x^* = \sum_{i=1}^{n} r_i x_i / \sum_{i=1}^{n} r_i \tag{1.1}$$

Biological Neural Networks
Konstantin V. Baev
© 1998 Birkhäuser Boston

This expression has the following interpretation: The main "attention" should be turned to the most reliable sources, which means to the more active channels. If a spike train is represented as a Paussion random process, the reliability of transferred information may be described as $1 - e^{-\lambda}$, where λ is the process intensity (frequency analog).

The preservation of the ability to work adequately after a partial break in information channels is peculiar to the brain, and has been designated as a holographic property. From the formal point of view, we will say that a mathematical function f, determined on a variable number of arguments, possesses the holographic property, if

$$f(n, x_1, \ldots, x_{i-1}, 0, x_{i+1}, \ldots, x_n) = \\ f(n-1, x_1, \ldots, x_{i-1}, x_{i+1}, \ldots, x_n), \qquad (1.2)$$

$$df(n, x_1, \ldots, x_n)/dx_i = 0, \text{ if } x_i = 0, \qquad (1.3)$$

where $i = 1, \ldots, n$. The simplest example of such a function is (1.1), provided $r_i(x_i) = 0$ and $dr_i(x_i)/dx_i = 0$, if $x_i = 0$. Indeed, when the reliability of ith value is equal to zero, the corresponding weight will be zero too. Thus there will be no systematic error, such as occurs in the case of usual averaging, when the weights are constant.

We have been considering until now the particular case of the full duplication of information in afferent channels. In general, the duplication may be only partial, and then the mathematical expectations of the corresponding random values cannot be regarded as equal. The main reason for this is the qualitative differences between information from receptors of different types. In addition, the thresholds of different receptors of the same type also differ.

Knowledge about correlations between sensory signals in different information channels is evidently somehow stored in the sensory information processing subsystem since this knowledge determines the functional organization of this subsystem. For the general case, it is convenient to present this knowledge in the form of conditional probability distributions.

Let $X = \{x_1, x_2, \ldots, x_n\}$ be the values of afferent signals received from different sources (including the internal model of the controlled object). Having designated x_i as the set of the signals from all sources except for i^{th}, we can write

$$P(X) = P(X_i \& x_i) = P(x_i)\, P(X_i/x_i) = P(X_i)\, P(x_i/X_i) \qquad (1.4)$$

In the case of $x_i = 0$, information from i^{th} source is unreliable, and therefore

$$P(X_i/x_i = 0) = P(X_i) \qquad (1.5)$$

Then, according to (1.4),

$$P(x_i = 0/X_i) = P(x_i = 0) \qquad (1.6)$$

(i.e., the conditional probability is equal to the absolute one). Finally,

$$P(X) = P(X_i \& x_i) = P(x_i = 0) P(X_i) \qquad (1.7)$$

which indicates that X_i and x_i are independent of each other. Knowledge about spatial correlations (these correlations depend, certainly, on the current state S of the controlled object) may be represented as conditional probability distribution $P(X/S)$. Having introduced an *a priori* probability distribution for S, we can obtain an *a posteriori* distribution $P(S/X)$ according to the well-known Bayes's formula:

$$P(S/X) = \frac{P(X/S)P(S)}{\int\limits_{\Omega Y} P(X/Y)P(Y)dY} \qquad (1.8)$$

where the denominator is equal to $P(X)$. If $x_i = 0$, then, similarly to (1.7),

$$P(X/S) = p(x_i = 0/S) \, p(X_i/S) \qquad (1.9)$$

Taking into account, that, as in (1.6), $P(x_i = 0/S) = P(x_i = 0)$, we can substitute (1.9) and (1.7) into (1.8). Then, after reducing the fraction, we obtain

$$P(S/X) = P(X_i/S) \, P(S)/P(X_i) = P(S/X_i) \qquad (1.10)$$

This means that broken channels are not taken into account, and therefore partial deafferentation does not cause systematic error.

It should be noted that the larger the dispersion of $P(x_i/S)$, the smaller the reliability coefficient corresponding to x_i. The reliability of an afferent information source thus depends on controlled object state, and so, naturally, does the amount of "attention" turned to the related afferent channel. In other words, the intrinsic model of the controlled object predicting its next state also controls attention function over internal feedback connections.

The optimum control value u must correspond to the most probable current state of the controlled object, which certainly is not always equal to the mathematical expectation. Thus,

$$u = u(S^*), \text{ where } S^* = \arg\left(\max_s P(S/X)\right) \qquad (1.11)$$

The model "sensory" signals X_M may be calculated according to S^* exactly in the same way.

It may be easily noted that the *argmax* function also "takes no acccount" of unreliable sources of information about $P(S/X)$, i.e., possesses the holographic property.

2 Functioning of the Internal Model of the Controlled Object

We introduced a formalized functional representation of the controlled object state in the form of an *a priori* distribution of its probability $P(S)$. If the intrinsic model of the controlled object does not include the object dynamics, $P(S)$ is time-independent and reflects the experience that the control system collected during its work. In more complicated cases, the control system can store different state probability distributions for different motor programs and recall an appropriate one from memory when a concrete motor program is started. In the latter case, the dispersion of $P(S)$ would obviously be less, and the current controlled object state could be determined more precisely. If a short-term memory exists and the model includes controlled object dynamics, the dispersion of $P(S) = P(S,t)$ may be significantly less (see Figure 70) and, consequently, the precision of determining the current phase state may be relatively high.

The main time constants of the model should appear the same as in the controlled object. However, the properties of the model dynamics must have one very important difference from those of the object. The object (mechanical system) itself cannot be moved instantaneously to any other phase state, because of its natural inertia. On the other hand, the model must allow

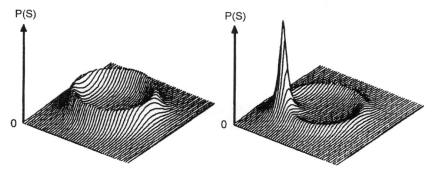

Figure 70—The density of the *a priori* probability distribution of the controlled object state, which the control system may "know" (in the context of a given rhythmic motor program). A, the case of absence of a short-term memory mechanism in the control system. The probability density is maximum everywhere along the optimal phase trajectory. B, the control system contains the model of controlled object dynamics. The peak of the probability density corresponds to the control system "idea" about the phase state of the controlled object at the current moment in time.

such a transition, to provide for the possibility of correcting a mistake in current state determination (i.e., to correct the internal representation of controlled object state in the control system). Let us consider the following example of functional description of a controlled object model.

Assume that controlled object dynamics is described by the following:

$$dS/dt = f(S,u) \qquad\qquad (1.12)$$

As stated above, the internal representation of the object state corresponds to the probability distribution density $P(S,t)$ (*a priori*, stored in the object model). If, at the time t the object was in the state S, at the next time $t+dt$, according to (1.12), it will be in the state $S+dS = S+f(S,u)dt$. Then,

$$P(S + dS,t + dt) = P(S,t) + P'_s(S,t)f(S,u)dt + P'_t(S,t)dt \quad (1.13)$$

where P'_s and P'_t are partial derivatives by state and time, respectively. Moreover, if there are no external sources of fluctuations,

$$P(S + dS,T + dt) = \hat{P}(S,t) \qquad\qquad (1.14)$$

where $\hat{P}(S,t)$ is an *a posteriori* evaluation of controlled object state probability distribution density, which corresponds to the current time moment. Then, the working of the internal model of the object dynamics can be described as a solution to the following differential equation:

$$P'_t(S,t) + P'_s(S,t) f(S,u) = (\hat{P}(S,t) - P(S,t)) / dt \qquad (1.15)$$

If there are no external fluctuations and errors in the working of the model, an *a priori* evaluation $\hat{P}(S,t)$ will not differ from $P(S,t)$, the right part of (1.15) will be equal to zero, and this equation will simply describe the object dynamics. Otherwise, a mismatch can be incurred between a prediction given by the model and the real controlled object behavior and, as a result of it, a break in the form of δ-function will take place in the right side of the equation. According to (1.15), an instantaneous sharp change in $P(S)$ occurs.

The equation (1.15) may be understood better after reading Chapter 4, where the principle of differential encoding of information is considered.

It is also possible to give a recursive representation of the working of the model. Taking into account that $P(S,t + dt) = P(S,t) + P'_t(S,t)dt$, from (1.15) we get:

$$P(S,t) + dt) = -P'_s(S,t)f(S,u)dt + \hat{P}(S,t) \qquad (1.16)$$

The functional schema of the spinal control system, which follows from all the abovementioned, is shown in Figure 71.

It should be emphasized that, in this case, we do not have the model of the desired result that is expected when the control task is finished. Instead, we use the functional model of the controlled object behavior (movement), which reflects the OCS's "experience" in dealing with the object. This proves that such a schema (in contrast to the schema of control according to a mismatch with a result) can really work after a break in the peripheral afferent feedback. It should be noted that at the same time we do have the case of control based on mismatch with a result, if the aim of control is assumed under such a result. The reader may well ask: "Can't a prediction given by the internal model also be regarded as the nearest thing to a result of control action? And if so, then what's the difference between the model of the controlled object and the model of a result?" We can answer this question in the following way. Both the concepts are of a functional type. The one that is applicable to a given informational signal is determined by the character of the information processing procedure. In the schema under consideration, a signal produced by the object model is processed as one of

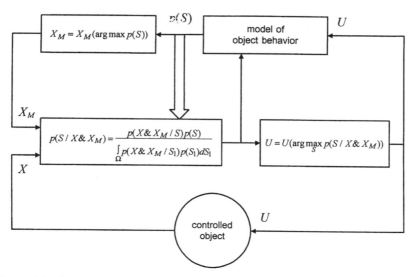

Figure 71—Schema of the functional organization of the spinal optimal control system. Detailed explanations are given in the text.

the peripheral afferent inflow components, instead of being compared with this inflow. We consider in Chapter 5 schemata describing such comparisons in a neural control system. Then, in Chapter 10 we show not only the way of using mismatch signals produced as the result of comparisons in the spinal OCS, but also describe an ascending tract conducting spikes of such a functional type.

3 The Spinal Optimal Motor Control System as a Neural Network

Let us consider how the probability approach may be used for function realization of the "encoder-decoder" type. The simplest neural structure for such a purpose is one layer of neurons without interconnections. Input signals $p_1, \ldots, p_m$ are transformed into output ones $q_1, \ldots, q_m$. In the simplest case the calculated function $y(x)$ has one argument, whose probability distribution is encoded by the input signals. The output signals encode the probability distribution of the function value. The input signal p_i (output signal q_i) is proportional to the probability of argument (function) value in i^{th} interval of its range. We will, meanwhile, consider the simplest case, when i^{th} and j^{th} intervals are not overlapping if $i \neq j$.

Thus, signals p_i and q_i can be represented mathematically as the differences of the corresponding probability distribution of the function values at the ends of ith interval: $p_i = F(x_{1i}) - F(x_{0i})$, $q_i = F(y_{1i}) - F(y_{0i})$. Let r_{ij} be the known conditional probabilities that if x belongs to i^{th} interval of its range, $y(x)$ belongs to j^{th} interval. Then,

$$q_j = \sum_{i=1}^{m} r_{ij}\, p_i, \quad j = 1, \ldots, n \tag{1.17}$$

In the general case, the conditional probabilities of negations of some events described in terms of argument value may be known. In our case, for example, s_{ij} may be the conditional probability of the fact that y belongs to j^{th} interval, if x does not belong to i^{th} interval. Then member $s_{ij}(1 - p_i) = s_{ij} + (-s_{ij})\, p_i$ appears in the corresponding sum instead of $r_{ij}\, p_i$. Then in the general case

$$q_j = \sum_{i=1}^{m_1} r_{ij}\, p_i - \sum_{i=m_1+1}^{m} s_{ij}\, p_i + \sum_{i=m_1+1}^{m} s_{ij}, \quad j = 1, \ldots, n \tag{1.18}$$

The calculation of the function $y(x)$ is represented in the form of the probability calculation of function value belonging to its range intervals. Such a way seems at first sight to be not very useful in practice. At least, it is very unlike the method of calculation in ordinary computers. To achieve high precision, it is necessary to divide the ranges of argument and function values into many intervals, which would require saving mn conditional probabilities. The amount of memory required for that may seem too wasteful for the calculation of one function only. However, the undeniable advantage of such a method of calculating functions is its universality. In addition, such a method is flexible enough to allow different degrees of precision in different regions of the argument range. But is such a method used in neural networks?

It is easy to notice that the three sums of (1.18) can be interpreted as the sum of excitatory influences, the sum of inhibitory influences, and the resting potential, respectively. Conditional probabilities r_{ij} and s_{ij} play the role of synaptic weights. Thus, we observe in this case a practically ideal correlation between the structure and the function of the simple neural network.

Let us suppose now that the events described by afferent signals can be compatible. In our case, this corresponds to partial overlapping of the argument variation intervals. A region occupied by two overlapping intervals A and B consists of three "simple" ones that do not overlap one another: $A \cap \overline{B}$, $A \cap B$, and $\overline{A} \cap B$. Knowing the probabilities of arguments belonging to the intervals A and B ($P(A)$ and $P(B)$, respectively), set by afferent signals, we can find analogous probabilities for "simple" intervals:

$$P(A \cap \overline{B}) = P(A)r(\overline{B}/A), P(\overline{A} \cap B) = P(B)r(\overline{A}/B),$$
$$P(A \cap B) = P(A)r(B/A) = P(B)r(A/B) \qquad (1.19)$$

There are two new distinctive features of the case under consideration, in comparison with the simpler case examined before. First, except for conditional probabilities of the "function value/argument value" type, one now has to take into consideration conditional probabilities of the "argument value 1/argument value 2" type (which are equal to zero, when there is no overlapping between corresponding intervals). Second, there are two ways of calculating the probability of an argument value belonging to the common part of the overlapping intervals. The overlapping of regions of the controlled object state that correspond to activating different receptors is typical of various sensory systems. Therefore, the analysis of neural mechanisms using such duplication during the processing of sensory information seems important.

Thus, we come back to the same problem of using the duplication of information in different channels for increasing its reliability. Now the difference between the *probability* of an event and the *reliability of a probability evaluation* becomes quite clear. On the basis of the analysis of functions calculated by a controlling neural network during sensory information processing, as described above, it is concluded that a simple functional model of a neuron can hardly be used for such processing. The main difficulty is that the synaptic weights of a simple neuron-like unit must be changed operatively according to the intensity of sensory signals.

Let us turn our attention to more complicated structures of synaptic links. The realization of multiplication in a function describing optimal control law requires complex synapses performing logical multiplication ("AND" operation). Synapses of this type can be observed in many brain sections (see Shepherd, 1983). The realization of the sensory information processing function (1.8) that require division is significantly more complex. It should be underlined that such functions can be, in principle, realized in the form of a network comprising simple neuron-like units. However, such a realization in practice turns out to be very complicated because a necessary functional property (specific nonlinearity) is not possessed by a single unit, and it has to be simulated by a rather large network. Nature has solved the problem of elaborating the necessary functional properties by complex synaptic organization. It has invented a new type of synapse and a new type of inhibition, called axo-axonal synapses and presynaptic inhibition, respectively. According to existing ideas (Schmidt 1971), presynaptic inhibition is based on the following mechanism. The burst of peripheral afferent activity evokes depolarization of primary afferent terminals *via* axo-axonal synapses with the aid of the system, which generates primary afferent depolarization (PAD). Depolarization of a terminal decreases the amplitude of the spikes in it, which leads to a reduction in the quantity of neurotransmitter released by a spike and, thereby, to corresponding attenuation of the sensory signal transferred through the terminal. The greater the depolarization of the terminal, the greater the attenuation coefficient.

The main distinction of presynaptic inhibition phenomenology is its comparatively large duration: more than 100 msec (Schmidt 1971). The idea presented above about the functional purpose of presynaptic inhibition enables us to give a natural interpretation of this property. In fact, if the presynaptic inhibition duration were as large as usual (postsynaptic) inhibition (about 10 msec), it would end shortly after the evoking spike and could not act on the subsequent spikes in other afferent terminals. The comparatively long duration of presynaptic inhibition allows decoding of the frequency code of the sensory signal, i.e., setting the dependence of the slowly altering depolarization level on spike frequency.

Our measurements revealed that the time constant in PAD development is equal to approximately 30–35 msec, which corresponds to the cut-off frequency of about 5 Hz, i.e., the constant level of presynaptic inhibition begins to grow significantly when the spike frequency becomes larger then 5 Hz. The cut-off frequency may be considered as a threshold value of spike frequency in an afferent fiber. If the spike frequency is greater than the cut-off one, the afferent signal will produce a significant effect on the PAD generating system. In other words, information about movement, when interspike intervals are more than 200 msec, would not be considered reliable enough. This is quite reasonable since the limb can perform a full cycle of, for example, scratching movement, during that time. This, therefore, shows that the neural subsystem calculating the attenuation coefficients of sensory information processing is relatively sensitive to the input signals.

The PAD generating system can also be considered as a system that calculates the reliability of sensory information. However, reliability (and probability) cannot be evaluated instantaneously. Such evaluation may be performed only on the basis of data about the sequence of spikes in a certain time interval. This requires, in turn, a simple variant of the memory mechanism, e.g., based on the fact that the time constant of this mechanism is considerably longer than the average interspike interval.

The second distinctive feature of the way the PAD generating system functions is a simultaneous development of PAD in all segments of the lumbosacral spinal cord enlargement (where the main part of the spinal control subsystem for hind limbs is localized) in response to stimulation of any bundle of afferent fibers entering only one (any one) of the enlargement segments. The functional purpose of this property is quite clear in the light of the idea that the spatial submodel of the controlled object uses the presynaptic inhibition mechanism for taking operative account of the reliability of sensory signals. Wide simultaneous distribution of PAD reflects the abovementioned existence of multiple spatial correlations in the peripheral afferents whose fibers enter different spinal cord segments.

It was shown above that the PAD generating system was activated not only by peripheral afferent signals, but also by central pattern generators, such as scratching and locomotor ones (Baev 1979, 1980; Baev and Kostyuk 1981, 1982), when peripheral afferent feedback was abolished. The functional purpose of centrally originated presynaptic inhibition of peripheral afferent signals can be fully understood only on the basis of the idea of the internal model of controlled object dynamics. Indeed, an intensive burst of afferent activity (from the internal model, or from the periphery) means the significant increase of the reliability of information transferred through the corresponding channels. Thus, according to (1.1), the resulting weights of other afferent flow components will be decreased. It is this type of attenuation that is provided by presynaptic inhibition through the PAD generating system.

It follows from the above-described concept of the functional role of PAD that (simply speaking) the value of depolarization of one afferent terminal, evoked by signals passing through another afferent fiber, corresponds to the conditional probability of a signal in the first afferent channel, if there is a signal in the second afferent channel. Consequently, activation of high-threshold afferents would, as a rule, evoke depolarization of the terminals of low-threshold afferents, but not vice versa. If the existence of such a phenomenon had not been proven experimentally (Schmidt 1971), it could be discovered theoretically! This example shows that our theoretical ideas can really claim to be more than mere explanations of obtained experimental data.

Abbreviations

AD	anterodorsal thalamic nucleus
AF	afferent flow
APA	arcuate premotor cortex
AV	anteroventral thalamic nucleus
CDP	cord dorsum potential
CM	centromedian nucleus
CMAd	caudal cingulate motor area on the dorsal bank
CMAr	rostral cingulate motor area
CMAv	caudal cingulate motor area on the ventral bank
CNS	central nervous system
CO	controlled object
COMT	catechol-O-methyltransferase
CPG	central pattern generator
CS	conditioned stimulus
DOPA	dihydroxyphenylalanine
DRP	dorsal root potential
DSCT	dorsal spinocerebellar tract
EO	effector organ
EPSP	excitatory postsynaptic potential
FRA	flexor reflex afferents
GABA	gamma-aminobutiric acid
GPe	external segment of globus pallidus
GPi	internal segment of globus pallidus
HC	highest centers
HLR	hypothalamic locomotor region
ic	informational context
icis	informational component of initiating signal
IPSP	inhibitory postsynaptic potential
is	initiating signal
LD	laterodorsal thalamic nucleus
LR	locomotor regions
LS	locomotor strip

M   motoneurons
MAO   monoamine oxidase
MC   motor cortex
M-cell   Mauthner cell
MD   mediodorsal thalamic nucleus
MLR   mesencephalic locomotor region
MPTP   1-methyl-4-phenyl-1,2,5,6-tetrahydropyridine
MRI   magnetic resonance imaging
NMDA   N-methyl-D-aspartate
NN   neural networks
NR   no rearrangement points
OCS   optimal control system
PAD   primary afferent depolarization
PET   positron emission tomography
PD   Parkinson's disease
PM   premotor area
RP   reordering plot
SMA   supplementary motor area
SNc   substantia nigra pars compacta
SNr   substantia nigra pars reticulata
SOCT   spino-olivo-cerebellar tract
SRCT   spinoreticulocerebellar tract
STN   subthalamic nucleus
US   unconditioned stimulus
VAmc   nucleus ventralis anterior pars magnocellularis
VApc   nucleus ventralis anterior pars parvocellularis
VLo   nucleus ventralis lateralis pars oralis
VM   ventromedial thalamic nucleus
VSCT   ventral spinocerebellar tract

Index

Page references in bold refer to illustrations.